Fiber Optics and Optical Isolators

By

Dr. Chris J. Georgopoulos
Professor of Electronics
Democritos University of Thrace
School of Engineering
Xanthi, Greece

© Copyright 1982
1st Edition
98 97 96 95 94 93 92 91   10 9 8 7 6

Don White Consultants, Inc.
State Route 625
P.O. Box D
Gainesville, Virginia 22065 USA
Telephone: (703) 347-0030
Telex: 89-9165 DWCI GAIV.

Library of Congress Catalog Card Number 81-52618

Printed in the United States of America

ACKNOWLEDGEMENT

I am indebted to many people for their advice, assistance, and suggestions during the development of this book. First, I wish to thank Mr. Donald R.J. White, President of Don White Consultants, Inc. (DWCI) for encouraging me to write this book and for his very useful suggestions during the preparation and editing of the manuscript. I wish, also, to thank Mr. Edward R. Price, Editor, Educational Materials Department of DWCI for bridging the long distance by his regular and punctual airmailing of useful and appropriate materials for the manuscript to me.

I would like to express my sincere appreciation to several individuals of the Engineering School of the University of Thrace, Greece, and especially in regard to graduate students Messrs. C. Koukourlis and A. Marras for their comments and help in improving the accuracy and the clarity of the text. I am also indebted to Mr. G. Vaidis and Mrs. H. Kondili for the careful preparation of the drawings and the special help rendered by Mrs. A. Anagnostou-Karamalidou for her skillful typing of the draft.

Besides the credit that is given in each case throughout this book for any material used, many thanks are herein extended to the editors of various publications and the authors of many excellent papers on fiber optics and related topics, who have granted permission for using excerpts from their published works.

Finally, I am grateful to my wife and two daughters, who not only tolerated my long hours of manuscript work, but also encouraged and assisted me in every possible way. Special thanks go to my daughter, Voula, who, although an undergraduate Electrical Engineering student, devoted a considerable part of her valuable time in proofreading the entire manuscript and offering numerous suggestions.

FOREWORD

It is with great pleasure on behalf of Don White Consultants, Inc. (DWCI) that I release *Fiber Optics and Optical Isolators*. This is another of our new handbooks on EMC and related topics to be published.

Each of our handbooks are prepared by recognized experts in their field. Several are now in preparation. DWCI's role is to provide the technical guidance, editing, logistics, financing, publishing and promotion. These books will provide a major contribution to the EMC and related technologies for years to come.

Regarding this handbook, *Fiber Optics and Optical Isolators*, it fills an existing void. This book has been prepared for engineers/designers as well as technicians who are engaged in fiber optics technology and its application in the electronics industry. This handbook also has been written in terms of the advanced state-of-the-art and carefully illustrated in such a manner that it can be used in tutorial and seminar courses, as well as at undergraduate levels of instruction. Therefore, it provides an invaluable design guide, and an adjunct to existing literature, dealing with this important and dynamic substantive area. The author, Dr. Chris J. Georgopoulos invites your comments. Similarly DWCI welcomes correspondence from the many readers who may wish to comment on any aspect of this book.

January 1982
Gainesville, Virginia USA

Donald R.J. White

ii

Handbooks Published by DWCI

(1) White, Donald R.J., *Electrical Filters—Synthesis, Design & Applications*, 1980.

(2) White, Donald R.J., Volume 1, *Electrical Noise and EMI Specifications*, 1971.

(3) White, Donald R.J., Volume 2, *Electromagnetic Interference Test Methods and Procedures*, 1980.

(4) White, Donald R.J., Volume 3, *Electromagnetic Interference Control Methods & Techniques*, 1973.

(5) White, Donald R.J., Volume 4, *Electromagnetic Interference Test Instrumentation Systems*, 1980.

(6) Duff, Dr. William G. and White, Donald R.J., Volume 5, Electromagnetic Interference *Prediction & Analysis Techniques*, 1972.

(7) Hill, James S. and White, Donald R.J., Volume 6, *Electromagnetic Interference Specifications, Standards & Regulations*, 1975.

(8) White, Donald R.J., *A Handbook on Electromagnetic Shielding Materials and Performance*, 1980.

(9) Duff, Dr. William G., *A Handbook on Mobile Communications*, 1980.

(10) White, Donald R.J., *EMI Control Methodology & Procedures*, 1982.

(11) White, Donald R.J., *EMI Control in the Design of Printed Circuit Boards and Backplanes*, 1982. (Also available in French.)

(12) Jansky, Donald M., *Spectrum Management Techniques*, 1977.

(13) Herman, John R., *Electromagnetic Ambients and Man-Made Noise, 1979*.

(14) *Hart, William C. and Malone, Edgar W., Lightning and Lightning Protection*, 1979.

(15) Kaiser, Dr. Bernhard E., *EMI Control in Aerospace Systems*, 1979.

(16) Feher, Dr. Kamilo, *Digital Modulation Techniques in an Interference Environment*, 1977.

(17) Gard, Michael F., *Electromagnetic Interference Control in Medical Electronics, 1979.*

(18) *Carstensen, Russell V., EMI Control in Boats and Ships*, 1979.

(19) Georgopoulos, Dr. Chris J., *Fiber Optics and Optical Isolators*, 1982.

(20) Mardiguian, Michel, *How to Control Electrical Noise*, 1983.

(21) Denny, Hugh W., *Grounding for Control of EMI*, 1983.

(22) Ghose, Rabindra N., *EMP Environment and System Hardness Design*, 1983.

(23) Mardiguian, Michel, *Interference Control in Computers and Microprocessor-Based Equipment*, 1984.

(24) *EMC Technology 1982 Anthology*

Notice

All of the books listed above are available for purchase from Don White Consultants, Inc., State Route 625, P.O. Box D, Gainesville, Virginia 22065 USA. Telephone: (703) 347-0030; Telex: 89-9165 DWCI GAIV.

PREFACE

Fiber Optics is a dynamic and rapidly expanding technology.
Over the past decade optical fiber loss has been improved by more
than an order of magnitude along with major advances in fiber bandwidth, strength and other characteristics. Performance of other components, such as emitters, detectors and connectors, also has been
greatly improved.

Optical Isolators have also been available commercially for about
a decade. They are widely used in a variety of applications such as
communications, data processing, process control, power supplies and
solid-state relays. In these instances they provide high-voltage
isolation interfaces, ground-loop isolation and other switching and
sensing functions.

For many systems, low-loss optical fibers offer significant advantages compared with metallic cables, which include wider bandwidth,
larger repeater spacing and smaller cable cross section. In addition,
optical transmission offers total electrical isolation, since optical
cables neither emit nor pick up electromagnetic radiation. Therefore,
the problems of RFI, EMI, ground loops, and sparking associated with
electrical cables can be eliminated. The overall trend is an exponentially increasing variety and number of applications as the advantages
of using fiber optics for these applications become more widely recognized.

This book provides a comprehensive presentation of fiber optics
and optical isolators with the emphasis placed on the fiber optics
field that has been thriving and shows signs of continuing that way.
Considerable effort has been made to present material at such a level
that there is consistent progression from concepts and design considerations to applications without getting mired in many theoretical
details. All circuits have been selected with great care to illus-
trate the tremendous range of applications where optical isolators
and fiber optics provide solutions to practical problems. It is
organized to facilitate rapid solutions to various problems confront-
ing the practicing engineer, and yet the level of presentation is
useful to technicians as well. Also, the book can be used in tutorial and seminar courses as well as at undergraduate level teaching.
A good portion of the material of this book is the outgrowth of the
author's notes prepared for an elective undergraduate course for

Electrical Engineering students in the Engineering School of the University of Thrace, Greece.

Chapter 1 of this book presents a brief review of some fundamental EMI and other related problems which occur in conventional hardwire coupling and suggests optical coupling techniques as alternative solutions. Chapter 2 deals with the phenomena of optical coupling and discusses various components such as optical sources, detectors and couplers. Chapter 3 is devoted to optical isolators, covering a wide range of applications as well as their capabilities and limitations. The properties of fiber cable preparation are presented in Chapter 4 along with a discussion of connecting elements. Chapters 5 and 6 discuss the fundamentals and applications of digital and analog fiber optic links, respectively, along with various selected circuits. Chapter 7 deals with the selection criteria for fiber optic links and associated components and touches upon the problem of standardization, while Chapter 8 describes some combined optical and hardwire coupling techniques in connection with electro-acousto-optic devices and wavelength multiplexing. The use of fiber optics in control and measurements is studied in Chapter 9. In Chapters 10 and 11, the current state-of-the-art of fiber optic components is reviewed and projected uses of fiber optics and market trends are detailed.

Xanthi, Greece
January 1982

Dr. Chris J. Georgopoulos
Professor

TABLE OF CONTENTS

FIBER OPTICS AND OPTICAL ISOLATORS

CHAPTER 1 AN INTRODUCTION TO OPTICALLY COUPLED AND FIBER OPTIC SYSTEMS

CHAPTER 2 PHENOMENA OF OPTICAL COUPLING - SOURCES AND DETECTORS

Table of Contents

CHAPTER 3   OPTICAL ISOLATORS

CHAPTER 4 FIBER OPTIC COUPLING AND CONNECTIONS

Table of Contents

CHAPTER 5   DIGITAL FIBER OPTIC LINKS

CHAPTER 6 ANALOG FIBER OPTIC LINKS

Table of Contents

CHAPTER 7   SELECTION CRITERIA FOR FIBER OPTIC LINKS AND ASSOCIATED COMPONENTS

CHAPTER 8 COMBINED OPTICAL HARDWIRE COUPLING

CHAPTER 9 FIBER OPTICS IN CONTROL AND MEASUREMENTS

CHAPTER 10   ASSESSMENT OF CURRENT CAPABILITIES

CHAPTER 11   FUTURE TRENDS

Table of Contents

ILLUSTRATIONS AND TABLES

FIBER OPTICS AND OPTICAL ISOLATORS

Illustrations

Illustrations

Illustrations

Illustrations

CHAPTER 6   ANALOG FIBER OPTIC LINKS

Illustrations

Illustrations & Tables

CHAPTER 10   ASSESSMENT OF CURRENT CAPABILITIES

CHAPTER 11   FUTURE TRENDS

TABLES

Tables

ABBREVIATIONS AND SYMBOLS

Å	Angstrom
a-c	Alternating Current
A.D	Audio Demodulator
AGC	Automatic Gain Control
A.M	Audio Modulator
AND	Logic Gate "AND"
APD	Avalanche Photo-Diode
AOIs	Acoustic-Optic Interaction Devices
BER	Bit Error Rate
byte	a single group of bits
β	Transistor current gain
BW	Bandwidth
CATV	Community Antenna Television
CC	Close Confinement
CC	Consultative Committee
CCITT	International Telegraph and Telephone
CMOS	Complementary Metal-Oxide Semiconductor
CMR	Common Mode Rejection
CTR	Coupling Transfer Ratio
CTR	Current Transfer Ratio
CVC	Current-to-Voltage Converter
CW	Continuous Wave
dB	decibel
d-c	Direct Current
DDS	Doped-Deposited Silica
DG	Differential Gain
DIP	Dual In-line Package
DIUs	Digital Input Units
DMUX	Demultiplexer
DP	Differential Phase
DPDT	Double-Pole Double Throw (Switch)
DPO	Digital Processing Oscilloscope
DTL	Diode Transistor Logic
ECL	Emitter-Coupled Logic
EEC	European Economic Community
EMI	Electromagnetic Interference
EMP	Electromagnetic Pulse

Abbreviations & Symbols

EMR	Electromagnetic Radiation
EO	Electro-Optic
EOTF	Electro-Optic Tunable Filter
FCC	Federal Communications Commission
FDM	Frequency Division Multiplexing
FET	Field Effect Transistor
FM	Frequency Modulation
FO	Fiber Optics
FSK	Frequency Shift Keying
F/V or FVC	Frequency to Voltage Conversion
G	Gain
GaAlAs	Gallium Aluminum Arsenide
GaAlAsSb	Gallium Alluminum Arsenic Antimonide
GaAs	Gallium Arsenide
GaInAs	Gallium Indium Arsenide
GaInAsP	Gallium Indium Arsenic Phosphide
GaSb	Gallium Antimonide
Gbits	Gigabits
Gbps	Gigabits per second
Ge	Germanium
GHz	Gigahertz (10^9 Hertz)
HE	Headend
Hz	Hertz
IB	Interface for Bus
IC	Integrated Circuit
IDP	Integrated Optical/Decoded Preamplifier
IEC	International Electrotechnical Commission
ILD	Injection Laser Diode
I/O	Input/Output
IR	Infrared
IVPO	Inside Vapor-Phase Oxidation
Kbits	Kilobits
KHz	Kilohertz
LD	Light Detector
LED	Light Emitting Diode
LOC	Large Optical Cavity
LOS	Local Origination Studio
mA	milliamp
Mbits	Megabits
MCVD	Modified Chemical Vapor Deposition
MDS	Minimum Detectable Signal
MHz	Megahertz (10^6 Hertz)
MOV	Metal-Oxide Varistor

Abbreviations & Symbols

MUX	Multiplexer
mV	millivolt
μm	micrometer (10^{-6} meter)
μP	microprocessor
μs	Microsecond (10^{-6} second)
μV	microvolt (10^{-6} Volt)
N.A.	Numerical Aperture
NC	Normally Closed
NEP	Noise Equivalent Power
NO	Normally Open
NRZ	Non Return-to-Zero
ns	nanosecond
ν	frequency (here optical)
OEM	Original Equipment Manufacturers
OFF	Cutoff State
ON	Conduction State
OR	Logic gate "OR"
OTDR	Optical Time Domain Reflectometer
OTF	Optical Transfer Function
OVPO	Outside Vapor-Phase Oxidation
PC	Printed Card
PCM	Pulse-Code Modulation
PD	Photo-Diode
p.f.	packing fraction
PIN or P-i-n	P-intrinsic-N Junction
PNP	Transistor "PNP" type
P-P	peak-to-peak
PPM	Pulse Position Modulation
PRM	Pulse Rate Modulation
PSK	Phase-Shift Keying
PUT	Programmable Unijunction Transistor
PUR	Polyurethane
PVC	Polyvinyl – Chloride
PWM	Pulse Width Modulation
RF	Radio Frequency
RFI	Radio Frequency Interference
RMS	Root Mean Square
SCR	Silicon Controlled Rectifier
SH	Single Heterojunction
S/N or SNR	Signal-to-Noise ratio
TDMA	Time Division Multiple Access
TP	Twisted Pair
TTL	Transistor-Transistor-Logic
TV	Television

Abbreviations & Symbols

UHF	Ultra High Frequency
UJT	Unijunction Transistor
VCO	Voltage-Controlled Oscillator
VDE	Verband Deutscher Electroniker
V/F or VFC	Voltage-to-Frequency Converter
VFH	Very High Frequency (30-300 MHz)
V_{SAT}	Saturation Voltage
VSWR	Voltage Standing Wave Ratio
VTR	Video Tape Recorder
WDM	Wavelength Division Multiplexing
Z_o	Characteristic impedance

CHAPTER 1

AN INTRODUCTION TO
OPTICALLY COUPLED AND FIBER OPTIC SYSTEMS

Fiber optic cables, which are made of glass or plastic, do not radiate nor are they susceptible to Electromagnetic Interference (EMI). This is a big advantage because it permits data to be transmitted without noise or induced error. Using fiber optics, it is possible to eliminate the common problem of crosstalk or crosscoupling and achieve electrical isolation approaching 100 dB between interfacing equipment along with data security.

1.1 INTRODUCTION

Before discussing optical coupling signals, as an alternative to hardwire coupling, it seems appropriate to review some fundamental EMI and related problems in conventional hardwire coupling and certain common solutions. Optical coupling is addressed later in this chapter. The phenomena of optical coupling are examined in Chapter 2, whereas Chapter 3 is devoted to optical isolators. The remaining material in this volume covers fiber optic coupling for both digital and analog signals, suggestions for special applications, selection criteria, and an assessment for current and future capabilities. Problems and limitations inherent to optical coupling and techniques for improvements are also studied, always keeping in mind the behavior of optocouplers and fiber optics when in an electromagnetic environment.

One of the prime considerations in modern electronic systems design is the problem of communication between subsystems. This engineering concern is justified, because, as electronic devices and systems become more sophisticated and more numerous, the airspace becomes increasingly polluted with the radiation they generate. This pollution is manifested as radio noise, electrical noise, or radio frequency interference (RFI) and is collectively referred to as electromagnetic interference (EMI). This interference from electromagnetic radiation can affect any type of electronic equipment, from the simplest radio or TV receiver to sophisticated computer systems, aircraft transponders and medical-analysis equipment. Regulations

in the US and other countries are becoming increasingly strict con-
cerning the EMI generated from various equipment and manufacturers
may be required to redesign equipment as a result of these rules.*

An electrical signal over a pair of lines requires a continuous
conduction path which is grounded at one end to avoid a floating
potential above some zero reference. Systems using coaxial conductors
have similar grounding requirements. Unless special connectors are
used, coaxial systems create multipoint grounds. Usually, floating
systems are used to avoid ground loops. Electronic systems directly
interconnected by these signals must share a common ground potential
to avoid erroneous signal generation by stray currents in grounding
loops. Although many ground isolation methods exist, the constant
threat of signal deterioration prevails whenever ground potential
differences can occur along the signal paths.

Computers, peripherals, and other digital devices are both
sources and receptors of EMI. They are sources because they contain
clock and other square-wave signals with many harmonic frequency components. As computing speed rises, the frequency spectrum of the
generated EMI broadens, and digital circuits have great potential for
interfering with other electronic equipment. EMI shielding of signal
lines has long been the only curative method. For system protection,
it has become good engineering practice to reduce EMI susceptibility
by restricting the bit-rate over shielded or coaxial cables to levels
much below the upper rate limits. In all such cases, signal protection against EMI is fundamentally dependent upon shielding integrity.

Signal ringing** in shielded, twisted pairs or in coaxial cable
is a constant hazard and can create serious engineering problems.
Good engineering practice requires signal pulse stretching and reduction in signal rise and fall times. These solutions, however, again
restrict the pulse-rate upper limit as a function of the signal-path
length.

In direct wire coupling, consideration must be given to electrical wires that originate from a common source or from point-to-point
connections, where a shorted conductor can induce failure in a sub-
system that would otherwise be able to operate in a degraded mode.
Other areas, where difficulties in signal transmission exist, include
those where there is possible danger to personnel or equipment,
directly or indirectly, as in cases of sparks in an explosive environment. In some instances, where metallic wires are used, security
problems may arise due to possible uncontrolled access to sensitive

---

* Some countries are considering legislating susceptibility.
** Ringing is defined here as a decaying oscillatory transient of
   current or voltage.

information, especially in military applications.

All previous problems can be solved in a simple, uncomplicated manner using optical couplers and/or fiber optics. Optoelectronic isolators and couplers are replacing the slow and bulky electromechanical devices that previously isolated delicate circuitry. Optical devices can prevent unwanted interaction among any physically separable parts internal to a system, as well as isolate a system from the outside world.

1.2 EMI AND RELATED PROBLEMS IN CONVENTIONAL COUPLING AND COMMON SOLUTIONS

Electromagnetic interference can create severe problems in solid-state logic circuitry. In the expanding use of solid-state logic in industry, science, and medicine it has been seen that many circuits fail to operate properly due to the proximity of high-energy equipment, such as motors, welders, pulsed lasers, and other machines that generate substantial amounts of EMI.

1.2.1 Sources of Noise and Regulations Regarding EMI

Electromagnetic interference is any unplanned electrical emission that impinges on an operating electronic equipment or system and causes performance degradation or malfunction. The extent of the noise encountered is not always anticipated when the circuit is first conceived. An example of this is when a test equipment manufacturer designs a general-purpose product that may find its way in to the vicinity of a plasma generator. This is generally viewed as an external interference, combining many possible sources, and with no high or low frequency limit. Among specific sources of external noise are electromechanical circuits that switch inductive loads, electric discharge devices, and RF generators.[1,2]

The equipment need not be massive to radiate at unacceptable levels. Even a small relay can cause trouble if it is placed in the vicinity of a highly susceptible circuit or component. Likewise, the device does not have to be operating at a particularly high average power level. For example, some types of equipment produce a high-energy discharge in a few microseconds and then remain passive for a long period. In a computer system, problems resulting from EMI can range from single bit errors to complete destruction of programs or files.

In the interest of maintaining a clean EMI environment as sources of EMI proliferate, the Federal Communications Commission (FCC) in the US and many regulatory agencies around the world are imposing progressively stricter rules governing EMI emissions from various

types of equipment.[3] At the present time, the most comprehensive
set of regulations regarding EMI emission limits has been established
by the Verband Deutscher Electroniker (VDE), an independent association of electrical engineers in West Germany that prepares regulations
and conducts tests for equipment safety, EMI emissions, et cetera.
Various countries have established their own regulations regarding
EMI, and the European Economic Community (EEC) is working toward a
common set of regulations for its member countries. Figure 1.1 shows
the permissible levels of on line-conducted interference and radiated
interference as adopted by the FCC and VDE.[4]

In designing equipment to meet EMI standards, two basic types of
EMI emissions are considered: conducted and radiated. Conducted emissions consist of interference conducted through the ac power line.
Radiated emissions consist of electromagnetic energy radiated from the
equipment and connecting cables. Thus, fundamentally, EMI can be
attacked in two ways: (1) prevent it from reaching susceptible circuits, or (2) keep it contained within shielded enclosures where it
can do no harm. In either case, the task requires eliminating the
means whereby EMI travels from a source to a point of disturbance by
radiation, conduction or a combination of both. As discussed in the
following subsections, special consideration must be given to ground
loops, safety, crosstalk, routing of wires, shielding and common-mode
and differential-mode rejection.

1.2.2 Ground Loops and Safety

Ground loops are created when a chassis, frame, or bus is used
as a common ground for two or more equipments. Because two ground
points do not have exactly the same potential, more than one ground
on a signal circuit or signal-cable shield creates a current flow between grounds. Even a fraction of a volt difference in two ground
potentials can cause an appreciable amount of current to flow through
a ground loop which, in turn, can produce excessive electrical noise
in any low-level circuitry.

The dotted connections in Fig. 1.2 show how two ground loops
are made by interconnecting equipments located in different buildings.
The first loop generates a current due to the potential difference
between the two building grounds. This current is in series with the
lower signal lead, causing ground-loop noise to be injected into the
transmission line. The second ground loop is formed by the signal-
cable shield from the signal source to the amplifier. The ground-
loop current in the shield is coupled to the signal pair through the
distributed capacitance of the signal cable.

There are several ways to avoid ground loops.[5,6] For example,
the ground loops in Fig. 1.2 can be eliminated by having the pulse
amplifier float and interrupting the ground of the shield at the load

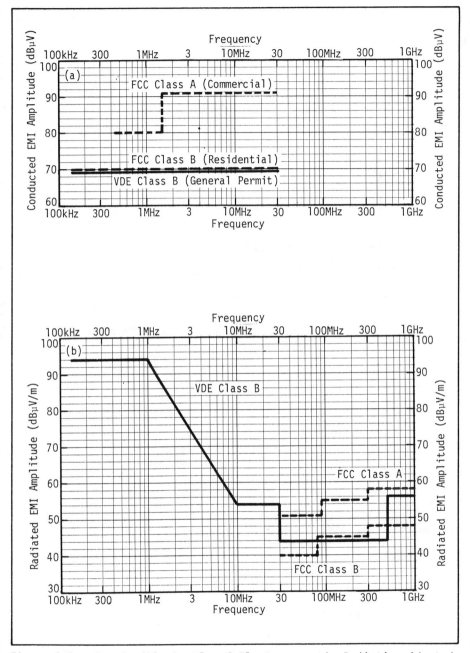

Figure 1.1 - Permissible Levels of Electromagnetic Radiation Adopted by FCC and VDE.

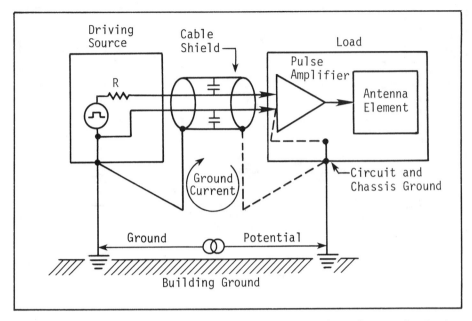

Figure 1.2 - Ground Loops Formed by Interconnecting Electronic Equipment Located in Different Buildings.

side. In all cases, signal-cable shields should be grounded only at one end.

Another popular method of interrupting ground loops is to create an open circuit between the source and the load, but in such a way that the circuit still transmits signals without permitting ground-loop currents to circulate. This can be done by using a transformer (Fig. 1.3a). This transformer breaks direct connection for ground-loop currents and transmits a-c signals and d-c pulses.

A better circuit that uses transformers to reduce noise is the balance drive (Fig. 1.3b). A transformer connected as a balun (BALanced to UNbalanced) can transmit steady-state signals as well as pulses, and greatly improves common-mode noise rejection (the ground noise between the sending and receiving ends of the line). Essentially, the balun presents high impedance to ground noise and low impedance to the useful signal.

The electrical codes in most countries, as well as good engineering practice, require that the chassis of all instruments in a system be at the same potential in order to prevent shock hazard to the operator. Usually, this is accomplished by connecting every chassis to the power line ground via a third wire in the power cord. The chassis of any instrument should never be allowed to float and power-line ground must not be removed or broken.

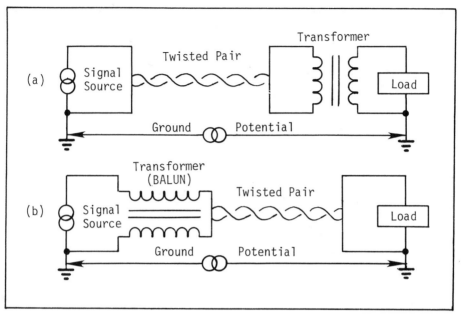

Figure 1.3 - Elimination of Ground Loops.

1.2.3 Crosstalk, Routing of Wires and Shielding

Current flowing in a wire or cable generates a magnetic field that surrounds the wire or cable. Conversely, a wire or cable may have current or voltage induced in it as a result of close proximity to an electromagnetic field. The signal introduced on a line by an adjacent active line is called crosstalk. For frequencies up to a few MHz, the effects of crosstalk signals can be studied by dividing them into electric and magnetic field effects. Figure 1.4 is the equivalent circuit of two transmission lines. Electric-field disturbances are a function of the line capacitances (C_ℓ) and mutual capacitance (C_m) between the disturbing and disturbed circuit. Magnetic field disturbances are a function of the line inductances (L_ℓ) and mutual inductance (L_m) between the circuits. For easier analysis these two disturbances can be considered separately.[7]

In the first case, the interfering voltage V_I couples through stray capacitance, C_c, to produce a voltage V_{ec} on the adjacent cable as shown in Fig. 1.5. The interfering cable and the adjacent cable have stray capacitances to ground (C_a and C_b). With cable load impedances high, the voltage on the inactive line will be:

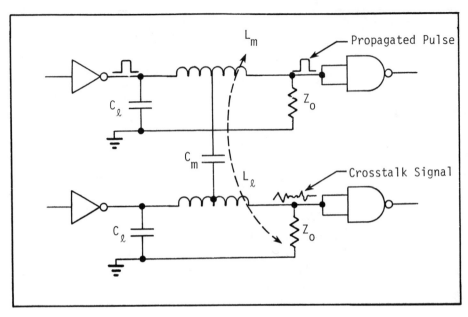

Figure 1.4 - Crosstalk Between Transmission Lines due to Electric and Magnetic Field Effects.

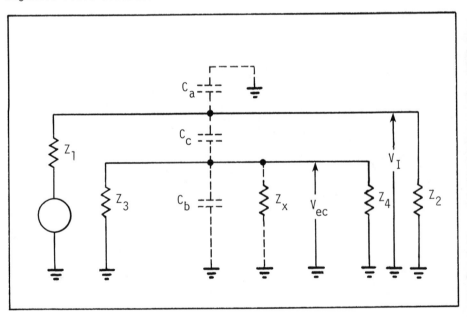

Figure 1.5 - Voltage Produced in a Cable Due to Capacitive Coupling.

$$V_{ec} = \frac{V_I \left[Z_x Z_b / (Z_x + Z_b) \right]}{Z_c + \left[Z_x Z_b / (Z_x + Z_b) \right]} \qquad (1.1)$$

If Z_x is a high resistance load (R_x), then:

$$V_{ec} = \frac{V_I C_c}{C_c + C_b} \sqrt{\frac{R_x^2}{R_x^2 + \left[1/2\pi f (C_c + C_b) \right]^2}} \qquad (1.2)$$

In the case of magnetic coupling, interference voltages are induced into a wire by flux likeages. The voltage induced in a loop (Fig. 1.6) by an adjacent wire of finite length carrying current is given by:

$$V_{mc} = (8.1 \times 10^{-8}) \, f \ell I \, \ln\left(\frac{d_2}{d_1}\right) \qquad (1.3)$$

where, V_{mc} = induced voltage in Volts

   f = frequency in Hz

   ℓ = loop length in cm

   I = current in Amp

d_1 and d_2 = loop distance in m

Electrostatic and electromagnetic coupling of noise into signal lines in the interface cables can be minimized by proper cable design and routing of wires during the installation. The interaction between two transmission lines decreases inversely with the square of the

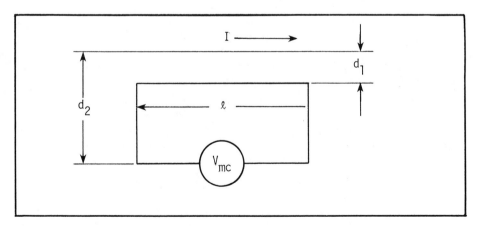

Figure 1.6 - Voltage Induced in a Loop by an Adjacent Current Carrying Wire of Length $>> \ell$.

separation distance of the two lines. If only a few lines are involved, it may be possible to reduce electromagnetic coupling by running the lines perpendicular to each other or, at least avoiding
parallel running.

It is also possible to use conductive coatings which can be
applied to non-conductive chassis and enclosures in order to provide
shielding against electromagnetic radiation (EMR) and electromagnetic
interference.[8]

The overall attenuation, or shielding effectiveness, is usually
expressed in dB as:

$$S_E = 20 \log_{10}\left(E_1/E_2\right) \qquad (1.4)$$

for the electric field,

$$S_H = 20 \log_{10}\left(H_1/H_2\right) \qquad (1.5)$$

for the magnetic field, and

$$S_T = 10 \log_{10}\left(P_1/P_2\right) \qquad (1.6)$$

for the total field power.

In these equations, E_1, H_1 and P_1 are the incident electric and
magnetic-field and power strengths, and E_2, H_2 and P_2 are the transmitted values. If the wave impinges perpendicularly (for maximum
penetration) to the surface of a large-area, planar, single-layer
shield, the total shielding effectiveness can be expressed in dB as:

$$S_T = A+R+B \qquad (1.7)$$

where,

> A = absorption attenuation
>
> R = reflection attenuation
>
> B = losses caused by internal reflections

In most cases, the B term can be neglected when A is greater than ten.
This happens in most cases at frequencies above the low end of the
audio range.[9]

Thus, with B left out, total field power in dB is:

$$S_T = A+R \qquad (1.8)$$

and A for homogeneous metals, can be calculated as follows:

$$A = 8.4 \times 10^{-2} t\sqrt{\mu f \sigma} \qquad (1.9)$$

where,

t = shield thickness in mm

μ = shield permeability in h/m

f = frequency in MHz

σ = shield conductivity in mhos/meter

The above calculations of shielding effectiveness assume a continuous metallic shield without electrical discontinuities. Generally, R is high for electric-field waves when the shield material has high conductivity. Materials with both high permeability and high conductivity generally provide the greatest shielding effectiveness; however, magnetic field wave shielding at low frequencies is difficult.

1.2.4 Common-Mode Rejection

Coaxial cables present good common-mode (ground noise) rejection. The twisted pair, unless it is shielded, is not good because adjacent objects can distort the propagation wave. Common-mode rejection in twisted pairs is greatly improved if the line has balanced drive as in Fig. 1.3. To further minimize crosstalk effects, a differential amplifier can be used to select, compare and amplify low-level signals in noisy environments. In the differential mode, unlike signals applied to the double-ended input result in an output proportional to their difference (Fig. 1.7a). In the common mode, like signals result in a negligible output. An interesting application is the use of differential amplifiers as line drivers and line receivers at the two ends of a transmission line (Fig. 1.7b). Crosstalk is significantly reduced because the twisted pair causes capacitive and inductive crosstalk to appear equally on both lines (common mode), hence enabling the line receiver to reject the undesired signals.

1.2.5 Other Problems and Solutions

In slow-speed systems operating in a high noise environment, high-noise-immunity logic or low-impedance circuits can be used that require higher voltage or current levels to create a detectable signal. In these cases, any noise coupled to an inactive line is only a small percentage of the total signal and far lower than the threshold levels of the high-voltage logic. However, this technique is not always practical, because considerable power is needed to transmit the high-voltage signals, especially if lines are terminated to avoid reflections. Active line termination may be the solution in connection with tri-state devices.[10]

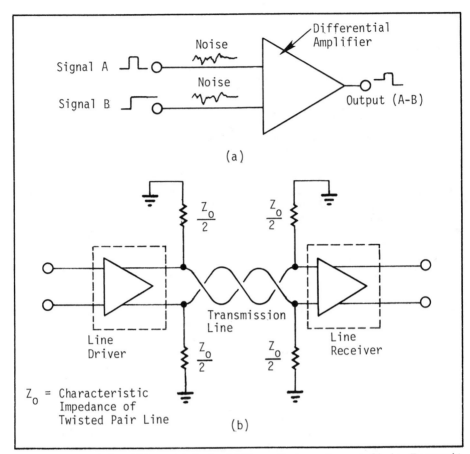

Figure 1.7 - (a) Differential Amplifiers Reduce Crosstalk by Transmitting the Useful Signal While Cancelling Noise Common to Both Inputs. (b) Amplifiers are Often Used at Both Ends of a Transmission Line in a Driver/Receiver Configuration.

Considerable thought must be given to selecting, locating, and connecting a power supply for a multiple load. Using a central power supply to supply the specified currents and voltages to a variety of remote loads is not always the best technique because, in certain systems, it may be necessary to bus a current of many thousand amperes to the electronic racks with a minimum d-c voltage drop.

It is important to understand the behavior of grounding circuits into the high frequency range, because high frequency noise sources can contain significant components. As one moves higher in frequency, the length of a ground path takes on significance beyond its resistance per unit length. The nature and quality of *ground* is considerably obscured as the length of the ground path becomes significant with respect to a wavelength.

Transient events, such as the discharge pulses coupled to the power lines by high powered lasers, may often contain significant components in the 100-MHz range. Arcing at the brushes of electric motors will frequently produce broad radio frequency noise into the 30-MHz region. Fluorescent lights, power hand tools, elevator motors, hood exhaust fans, the ubiquitous laboratory refrigerator, laser power supplies, RF generators used to modulate lasers or excite plasmas are but a few of the possible sources of noise which must be considered. Even though signals of interest to the electrochemical experimenter are relatively low in frequency, indeed, mostly dc, the possibility of interference at significantly higher frequencies must be considered.[2]

For general purpose transient protection, two types of components are often used at the semiconductor-device and system level: (1) metal-oxide varistors (MOVs) and (2) avalanche-diode voltage clamps.

Power line filters are available which may be plugged into the power circuit of the instruments in an experiment. Isolation transformers are effective as well in rejecting interference that may be carried along the power lines, but in general, the propagation path for radio frequency interference (RFI) is not easy to identify. For that reason, it is best to treat interference at its source whenever possible.

1.3 OPTICAL COUPLING AS AN ALTERNATIVE SOLUTION

Optical isolators and fiber optics are now well-proven technologies for transmission of intelligence in both digital and analog forms. They can protect delicate circuits from common-mode voltages, ground-potential differences and accidental overvoltages, while breaking ground loops.

1.3.1 Isolating Optoelectronic Pairs

An optical isolator, also called an optical coupler or an optically coupled isolator, is a device that converts an electrical signal at its input via a light source into photon energy. It then optically couples this energy to a photosensor to produce at its output a replica of the input. Modern semiconductor optical isolators use the light emitting diode (LED) as a light source. The LED is usually made of gallium-arsenide, operates in the near infrared region, and is optically coupled to a photosensing junction. This junction may be part of a photodiode, a phototransistor, or a photo-Silicon Controlled Rectifier (SCR).

For applications requiring a high frequency response, a p-i-n photodiode can be used, but its low sensitivity should be taken into account. If speed can be sacrificed for higher sensitivity, then a phototransistor is the one to be used in regular or Darlington configuration, the latter offering higher sensitivity. If speed cannot be sacrificed, an avalanche photodiode can be used which combines both high sensitivity and speed; however, it is more expensive than the previous devices. For applications requiring latching capabilities, a photo-SCR must be used.[11]

Optical isolators are used successfully in many applications to isolate voltage levels between circuits, prevent interference, eliminate dc ground loops, insulate people from the hazards of high voltage, amplify or attenuate signals and perform on/off switching.[12]

Figure 1.8 shows the use of optical isolators as line receivers with twisted pair (TP) lines in a computer-to-peripheral interface application, where they reduce the amount of ground loop current resulting from the effects of common-mode voltage. More details on this type of coupling and its limitations are given in Chap. 2.

In another application, shown in Fig. 1.9, optical isolators and twisted pair lines constitute one of the possible alternative solutions in motor control network, replacing mechanical parts, eliminating ground loops and protecting the logic circuits against the high-voltage transients from the load.[13]

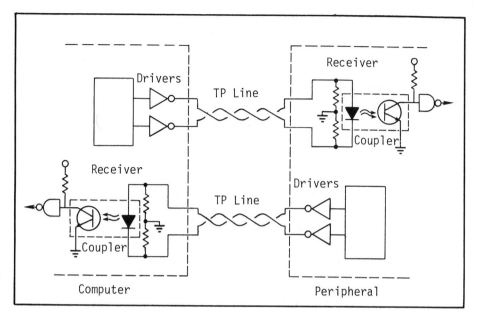

Figure 1.8 - Optical Isolators in Computer-to-Peripheral Interface.

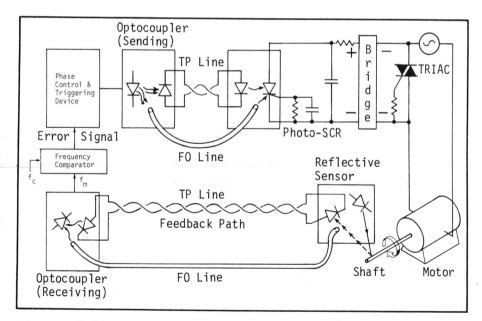

Figure 1.9 - Motor Speed Control System with Optocouplers and Twisted Pairs (TP) or Fiber Optic (FO) Lines in the Control Loop.

1.3.2 Fiber Optic Cables for Signal Transmission

The use of fiber optic data links has grown significantly during recent years. This strong interest arises from the advantages that optical fibers have over metallic conductors for transmitting information. The advantages depend on the application, but include the following:[14]

- Insensitivity to electromagnetic and radio-frequency interference. (note: Fiber optics may be susceptible to EMP gamma radiation unless special techniques have been used during their manufacturing).

- Immunity to ground-loop problems.

- Improved security compared with electronic cabling (no cross-talk among parallel cables).

- Elimination of combustion or sparks caused by short circuits.

- Flexibility in upgrading system capacity without need to install new cables.

- Low transmission loss (greater distance between repeaters).

- Wide transmission bandwidth.

- Potential low cost.

- Relatively small size, light weight, high strength, and flexibility.

- Suitability for digital communications and pulse modulation methods (fiber optic cable losses are independent of transmission frequency).

- Fewer government regulatory difficulties (because of the elimination of frequency allocation).

- Suitability for relatively high temperature.

The rapid growth of fiber optic technology has led to a situation in which commercial applications in the near future are likely to be limited by the ability of potential users and suppliers to understand and apply the technology and to establish standards. Practical fiber optic cable structures and electronic subsystems have been designed and numerous experimental systems of varying complexity have been installed in the field.

Fiber optic data links are becoming more practical because of recent developments in solid-state emitters. They now produce considerable speed-power products in the infrared portion of the frequency spectrum (around 3×10^5 GHz or a 1-μm wavelength), where fiber optic links usually operate. Emitters can be either lasers or light-emitting diodes. Along with the improved performance of emitters, detectors are also getting better. The silicon p-i-n detector is the most popular choice because it is the best match to the preferred optical frequencies of 0.7 to 1.1 μm. Light emitting diodes (LEDs) are incoherent and cause dispersion in the optical fibers. Thus, their bandwidth is limited. On the other hand, diode lasers are coherent sources with high bandwidth, but they are more expensive.

In Fig. 1.10 it is clear that a fiber optic system provides a large physical separation between the nodes of the common-mode voltage, e_{CM}, and is therefore superior to optical couplers in dealing with EMI.

Following are some representative applications of fiber optics, including data transmission and instrumentation.[15,16] Figure 1.11a shows a fiber optic system for measurements in high-intensity magnetic and electric fields. The point of measurement can be located remotely from the detector/electronics without suffering performance degradation. By keeping the sensor in the electronics box, adjacent to the preamplifier and shielded from outside electromagnetic interference, while bringing in light signals through fiber-optic cables, maximum sensitivity can be achieved. Figure 1.11b depicts a biomedical application with a fiber optic cable replacing conventional wires that could pose a danger for the patient due to high voltage apparatus operating at the monitoring end and due to possible faulty grounding. A digital data transmission interface between a computer main

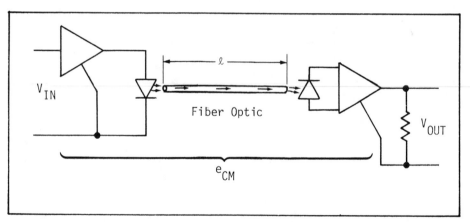

Figure 1.10 - Fiber Optic Data Link Arrangement Showing Common-Mode Node Separation.

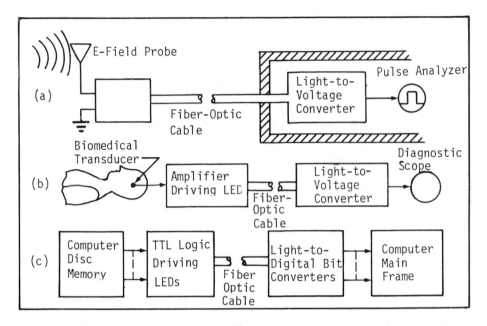

Figure 1.11 - Representative Applications of Instrumentation and Data Transmission Systems with Fiber Optic Cables.

frame and the disc memory is shown in Fig. 1.11c. A simple fiber optic or a bundle of fiber optics may be used depending on the type of transmission. In Fig. 1.9 fiber optic cables again constitute an alternative coupling solution in a high-voltage switching environment.

As a case history, it is also worth mentioning here an application relating to high energy physics, where several banks of capacitors, charged to thousands of volts, had to be sequentially discharged into a small loop of heavy wire to produce an ultra-high-strength magnetic field.[17] The extremely high-field intensity produced when any one bank of capacitors discharged, in addition to the close physical proximity of the separate capacitors, caused very serious spurious triggering problems until fiber optics came on the scene.

As can be seen from Fig. 1.12, a pulse produced by means of another photo-diode was available when the discharge of a particular bank occured. Several such channels allowed the proper time intervals between discharges to be accurately controlled.

Prior to the fiber-optic approach, induced levels of three to five volts on a double-shielded, 50-ohm transmission line occured whenever any one bank of capacitors discharged, regardless of grounding, loading, etc., and proper synchronization of discharges was

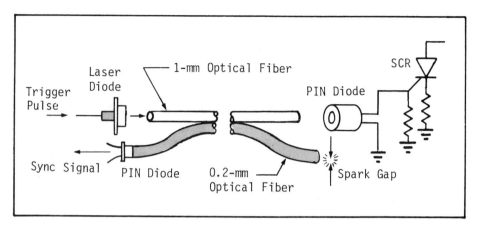

Figure 1.12 - A Laser Driven Scheme with Fiber Optic Cables for High-Voltage Discharge Applications (after Ref. 17).

almost impossible to achieve. With the fiber-optics system, no false triggering occurred at all.[17]

   It should be noted, however, that in solving interference prob-lems with fiber optics, there is a potential EMI problem in every optical link. The photodetector and associated high gain, low-noise amplifier need to be enclosed in a Faraday cage (a metal enclosure). The opening through which light enters the receiver needs to be very small with respect to the wavelength of the highest frequency RF fields to be encountered. This is an area that needs more attention and it will be covered in some detail in Chap. 2.

1.4 REFERENCES

1. White, D.R.J.; *Electromagnetic Interference Control Methods and Techniques*, Volume 3, Don White Consultants, Inc., 1973.

2. Hill, J.S. and White, D.R.J.; *Electromagnetic Interference Specifications, Standards & Regulations*, Don White Consultants, Inc., 1975.

3. Wong, D.T.Y.; *Controlling Electromagnetic Interference Generated by a Computer System*, Hewlett-Packard Journal, p. 17-19, September 1979.

4. Santoni, A.; *FCC's Computing-Equipment EMI Standards Pose Threat to Other Products' Shipments*; EDN, pp. 47-54, March 5, 1980.

5. Georgopoulos, C.J.; *Squelching Noise in Instrument Systems*, Machine Design, Vol. 43, No. 18, pp. 74-79, July 22, 1971.

6. Oliver, F.J.; *Practical Instrumentation Transducers*; Chapter 17, Hayden, N.Y., 1971.

7. Interference Technology Engineers' Master; *Cables and Connectors*; pp. 94-95, R&B Enterprises, Pa., 1979.

8. McDermott, J.; *EMI Shielding and Protective Components*; EDN, pp. 165-176, September 5, 1979.

9. Rashkow, B.; *An EMI Shield is Only as Effective as its Weakest Link - the Gasket*; Electronic Design 9, pp. 88-94, April 26, 1979.

10. Georgopoulos, C.J.; *DC Noise Margin Improvement in TTL Bus Organized Data Transmission Systems*; Proceedings of 1977 IEEE International Symposium on Circuits and Systems, Phoenix, Arizona, pp. 110-113, April 25-27, 1977.

11. Otsuka, W.; *Increased Reliability and Higher Gain Spurs New Optoisolator Applications*; Electronic Design, No. 26, pp. 92-95, December 20, 1978.

12. Hewlett-Packard Optoelectronics Division; *Optoelectronics Applications Manual*; McGraw-Hill, N.Y., 1977.

13. Georgopoulos, C.J., Athanasiadis, N., and Papamiltiadis, M.; *Optocoupling Devices to the Service of Electrical Machines*; International Conference on Electrical Machines, Sept. 15-17, 1980, Athens, Greece.

14.   Fagenbaum, J.; *Optical Systems: A Review;* IEEE Spectrum, p. 70, October 1979.

15.   McDermott, J.; *Fiber Optics Finding Grounding Use in Data Systems - and it Pays Handsomely;* Electronic Design 4, pp. 26-29, February 17, 1972.

16.   Wendland, P.H.; *Fiber Optics: At Work in New Instrumentation;* Electro-Optical Systems Design, pp. 35-37, April, 1980.

17.   Math, I.; *Optical Fiber Data Links Cover Diverse Applications;* Communications News, September, 1979.

CHAPTER 2

PHENOMENA OF OPTICAL
COUPLING - SOURCES AND DETECTORS

This chapter deals with the phenomena of optical coupling via air with very small and moderate separations between source and detector. Coupling through optical fibers is addressed in Chap. 4.

2.1 INTRODUCTION

Although light and its characteristics are familiar, light combined with electronics to perform a task is a relatively recent development. Light is radiant energy which can be generated, focused, transmitted, absorbed and detected in a variety of ways. It is easy to use light. Simple lenses, sources, and detectors, while small, are considered large when compared to the micrometer wavelength of electromagnetic radiation. Because of the small wavelength, beams can be collimated with low cost lenses.[1]

Light, while easily contained and controlled, must be transmitted to be used. Coupling of radiant energy takes place through many media. Air and fiber optics are the most common media for data transmission, control and communication purposes.

2.2 RADIOMETRIC AND PHOTOMETRIC DEFINITIONS, QUANTITIES AND UNITS

Geometrical relationships are of prime concern in the design of electro-optic systems. In this section some of the basic definitions are reviewed. More details can be found in References, Sec. 2.11, 2-4.

2.2.1 Electromagnetic Spectrum

Light is a form of electromagnetic energy and occupies a part of the spectrum of electromagnetic radiation. The distinction between the various radiations is primarily energy, which is proportional to frequency. Today, the characteristics of the relatively narrow optical band within the electromagnetic spectrum, coupled with quantum electronic have brought engineering to the era of optoelectronics.

2.2.2 Spectral Relationships [3]

Photometry deals with flux (in lumens) at wavelengths that are visible. Therefore, the unit symbols have the subscript v and the unit names have the prefix *luminous*. *Radiometry* deals with flux (in watts) at all wavelengths of radiant energy. Therefore, the unit symbols have the subscript e and the unit names have the prefix *radiant*.

From a geometrical standpoint, there are five generic terms and symbols as shown in Table 2.1. In this table the term *solid angle* is included for clarification purposes. Adding subscripts and prefixes quantifies these terms as radiometric or photometric units. Except for the difference in units of flux, radiometric and photometric units are identical in their geometrical concepts.

2.2.3 Optoelectronic Devices

Optoelectronic devices, which make use of the mutual interaction of radiation and the electronic structure of materials, have become widely used in recent years. Few areas of electronics have grown so rapidly and in so many dimensions as optoelectronic devices. This is mainly due to the improved manufacturing techniques developed for both sensitive silicon-based photodetectors and for gallium arsenide and gallium phosphide-based light emitters. It is also due to similar improvements that have made light-emitting diode display units and optoelectronic coupling devices more widely used.[4]

Table 2.1 - Generic Terms, Symbols and Geometrical Relationships in
          Optical Devices.*

Geometry	Term, Definition	Symbol	Defining Equation
	FLUX, rate of flow of energy, Q Q_e-radiant, Q_v-luminous, Q_q-photon.	ϕ	$\dfrac{dQ}{dt}$
	INCIDENCE, flux per unit area on a receiving surface.	E	$\dfrac{d\phi}{dA}$
	EXITANCE, flux per unit area from an emitting surface.	M	$\dfrac{d\phi}{dA}$
	INTENSITY, flux per unit solid angle from a remote source.	I	$\dfrac{d\phi}{d\omega}$
	STERANCE, flux per unit solid angle per unit area of emitting surface at angle θ with respect to surface normal.	L	$\dfrac{dl}{dA\cos\theta}$ $\dfrac{d^2\phi}{d\omega(dA\cos\theta)}$
	SOLID ANGLE	ω	$\omega = A/r^2$

* Adapted from Reference 3

   Although optoelectronic components are already extensively used
in control and regulation, data processing, and industrial and consumer electronics, continuing technological improvements will lead to
even more widespread applications.

   Since light emitting diodes, injection laser diodes and photodetectors are the essential devices used in fiber optic systems, they
will be examined in some detail in the sections that immediately
follow.

2.3 INCOHERENT OPTICAL SOURCES - LEDs

Incoherent optical sources, used in various applications, are usually one of four basic types: (1) *incandescent,* (2) *gas discharge,* (3) *electroluminescent,* and (4) *light emitting diodes (LEDs).* Of these, the light emitting diodes are relatively new. They came into existence in 1968 and today are most attractive to circuit designers because they exhibit long life like that of other semiconductors. All other light sources exhibit burn-out or half-life problems that are not easily solved by the user.

Light emitting diodes are simple to construct, extremely easy to modulate and have well-defined reliability/degradation characteristics. These characteristics make them particularly useful in system designs where relatively broad emission linewidth can be tolerated.

2.3.1 Radiation and Efficiency

Recombination radiation has been observed over a wide range of efficiencies in most of those semiconductors in which p-n junctions can be fabricated.[2] When the electro-luminescent radiation is incoherent, the diode source is described as a light-emitting diode (LED) type; when the radiation is coherent, the source is described as a laser type (see Sec. 2.4).

Looking at the possible applications for fiber optic communications and their performance requirements, it is obvious that LEDs are suitable light sources; in fact, very few of the systems installed to date use the laser. By far the largest percentage of use is one or another form of LED.

The optical link designer is interested primarily in the following parameters of the LED; (1) light output power, P, and (2) radiant intensity, J, i.e., the power per unit solid angle, θ. The radiant intensity is given by:

$$J = \frac{4P}{\pi\theta^2} \tag{2.1}$$

Manufacturers normally specify the quantum efficiency, as η, which relates light output to the diode current (Fig. 2.1)

$$\eta = \frac{\text{light emitted in photons}}{\text{diode current in electrons}} \tag{2.2}$$

Each photon has an energy,

$$E = h\upsilon = h\frac{c}{\lambda} \tag{2.3}$$

2.4

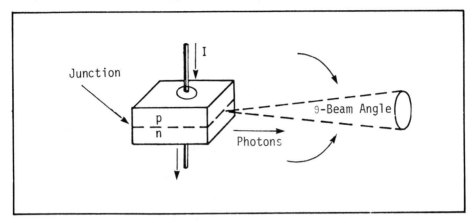

Figure 2.1 - Light - Emitting Diode P-N Junction Emits Photons While Diode Current Modulates Photon Emission.

A light power output, P, consists of $P/E = P\lambda/hc$ photons/sec and a diode current, I, consists of I/e electrons/sec.

then,
$$\eta = \frac{P\lambda e}{Ihc} \qquad (2.4)$$

or,
$$\eta = 0.857 \ (P\lambda/I) \qquad (2.5)$$

2.3.2 Material Characteristics and Types of LEDs

Incoherent or spontaneous emission devices are available covering a broader range of the spectrum than that of the laser diodes.[2,5] Table 2.2 shows a partial listing of LED sources and some of their characteristics[6] for use in optical fiber communications.

Basically, there are two configurations of LEDs[7,8] for optical communications: (1) *Burrus-type surface emitters* and (2) *edge emitters* as shown in Fig. 2.2. The characteristics of the first type (Fig. 2.2a) are high radiance, low thermal impedances and good coupling to fibers.[9] To effectively couple light, a lensed LED or a lensed fiber is usually used. The second type, the *edge emitter* diode (Fig. 2.2b) is usually a double heterojunction structure,[10] where guiding layers channel the light toward the fiber core from the side to form a narrower beam source than that of a surface emitting diode.

Basic drive circuits for diode light sources are rather simple, and are discussed in Chap. 3 and 5 of this handbook.

2.5

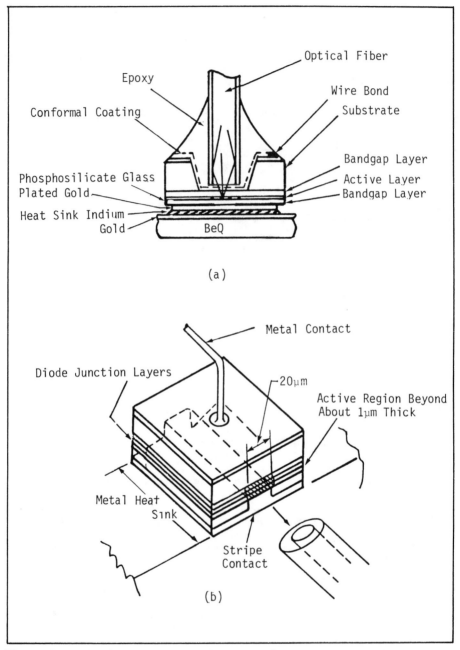

Figure 2.2 - Basic Structures of LEDs.   a) Burrus-Type Surface
Emitter.   b) Edge-Emitting Source.

Table 2.2 - Comparison of LED Material Characteristics

	Peak Emission Wavelength	External Efficiency	Rise Time
GaAlAs	0.82 to 0.84µ	2%	<10ns
GaSa (Si doped)	0.94µ	10%	400ns
GaAs (Zn doped)	0.90µ	1%	20ns
GaInAs	1.06µ	0.25%	5ns
GaInAsP	1.06µ	2%	5ns
GaInAsP	1.2µ to 1.3µ	3%	5ns

2.3.3 LED Arrays

Arrays of LEDs can be made to the user's specifications for both physical and optical properties. Color outputs ranging from green to far infrared are also available. LEDs can be fabricated in miniature arrays with very narrow cell-to-cell spacings. Focused or defocused output can be achieved by easily molded plastic lenses.[11]

Figure 2.3a shows a cross-sectional drawing of a diode structure with a spherical lens in the array. In this structure, the output power is emitted perpendicular to the junction plane. A photograph of a fabricated LED array[12] with self-aligned sphere lenses is shown in Fig. 2.3b. The spacing of the diodes is 0.5 mm; dimensions of the array chip are 5.0 mm in length, 0.6 mm in width, and 0.2 mm in height.

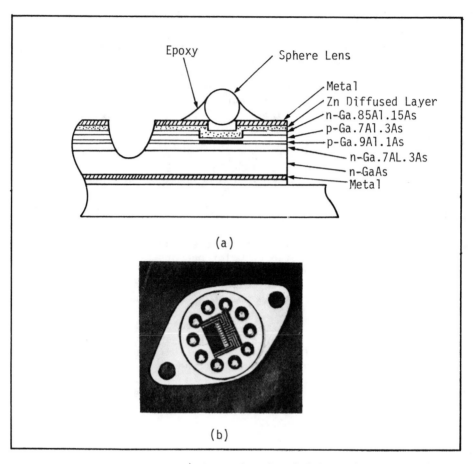

(a)

(b)

Figure 2.3 - LED Arrays. a) Cross-Sectional Schematic of a Diode
Structure in the Array. b) Photograph of Monolithic LED Array
(After Ref. 12).

2.4 COHERENT SOURCES - INJECTION LASER DIODES

A practical coherent source, such as a laser diode, has three basic requirements: (1) it must be constructed from a direct bandgap semiconductor; (2) it must have a Fabry-Perot cavity, consisting of two mirror-like surfaces defining the direction of photon flux; and (3) it must have a region formed which confines the radiation and the injected carriers.

2.4.1 Some Basic Structures of Laser Diodes

The Fabry-Perot cavity is generally created by cleaving the material along parallel crystal planes either of which may be coated, to a final reflectivity R_1 and R_2, as shown in Fig. 2.4.

The physical structure of a diode laser consists of a rectangularly shaped semiconductor die in which the radiant output is emitted from the edges of the diode in the recombination region of the junction.[2] The lateral size of the radiation area is usually defined by etching an opening in an oxide insulating layer and forming an ohmic contact region. This forms an *edge emitter* using *stripe geometry*.

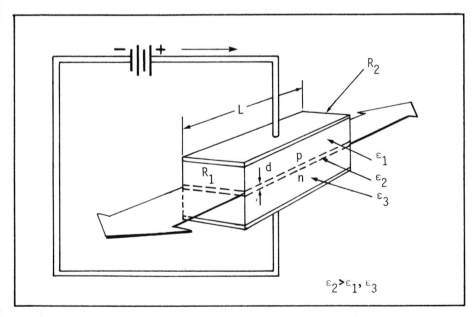

Figure 2.4 - Diagram Including Some Essential Features of a P-N Junction Laser: Fabry-Perot Reflectivities R_1 and R_2, Dielectric Waveguide d formed by Dielectric Constant $\varepsilon_2 > \varepsilon_1$, ε_3.

In order for lasing to take place, two conditions must be satisfied. A sufficiently high concentration of carriers must be injected into the active region to induce population inversion. Hence, the term *injection laser*. This is effectively accomplished by adopting a double heterojunction structure, in which a thin active layer, either n- or p-type $Al_yGa_{1-y}As$ (in the case of an AlGaAs laser), is sandwiched between n- and p-type AlGaAs layers. The other condition for laser oscillation is optical feedback, supplied by a pair of mirror facets at the ends of the device.

Well suited for use with fiber-optic materials, the stripe geometry enables the size of the radiating region to be matched to the size of the fiber by controlling the stripe width. Coupling loss can thus be minimized. Stripe width is generally about 1-1.3 μm. Figure 2.5 shows the basic stripe geometry of an early type double heterostructure laser, and gives typical dimensions of the layers.[13]

Of all types of lasers, the semiconductor injection laser diode (ILD) is the one exceptionally well-suited for fiber optic transmission. Some of the desired features are: it is physically small, inherently rugged, highly efficient and can be pumped and modulated simply and directly by means of the injected current.

For certain applications, bare laser diodes may be required. Such laser diodes are available mounted on a heat sink, and operate in any of the wavelength ranges shown in Table 2.3.[2,13]

Table 2.3 - Some Construction Materials and Wavelength Ranges for Laser Diodes.[2]

LAYER	WAVELENGTH RANGE	
	0.76-0.90 μm	1.1-1.5 μm
Contact	GaAs	$In_{1-w}Ga_wAs_zP_{1-z}$
Cladding	$Ga_{1-x}Al_xAs$	InP
Active	$Ga_{1-y}Al_yAs$	$In_{1-w}Ga_wAs_zP_{1-z}$
Substrate	GaAs	InP

2.4.2 Characteristics of Injection Laser Diodes [14]

A very useful characteristic of a diode laser is the curve of output light power as a function of drive current (P_o-I). The curves (Fig. 2.6) for a GTE Laboratories' diode laser, at various heat-sink temperatures, show the threshold current at which lasing (oscillation) begins (I_{th}) and the variation of the threshold current with

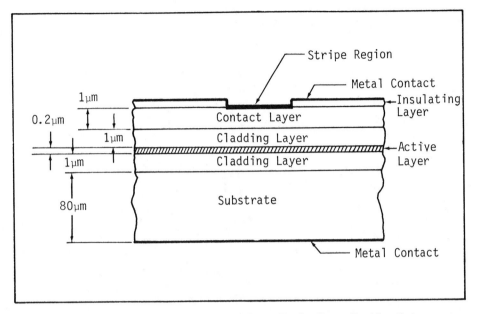

Figure 2.5 - Basic Stripe Geometry of an Early Type Double Hetero-structure Laser.

temperature. The pulsed P_O-I curve shows that threshold current de-creases with the device's temperature during pulsed operation. I_{th} is considered to be the point at which the tangent to the P_O-I curve intersects the current axis.

Below the threshold, the light output is a broadband, LED-type, spontaneous emission. Above the threshold, the lasing emission of the diode laser has the narrow bandwidth (<1nm) characteristic of an optical oscillator.

The threshold current increases nonlinearly with temperature, T, and follows the relationship,

$$I_{th} = I_A \exp\left(\frac{T-T_A}{T_O}\right) \qquad (2.6)$$

Here, I_A is the threshold current at an ambient temperature T_A (^{O}K), and T_O is a constant for the device, called the characteristic temper-ature. The value of T_O is a measure of the temperature sensitivity of I_{th} and falls in the 130 to 160-K range for AlGaAs devices.

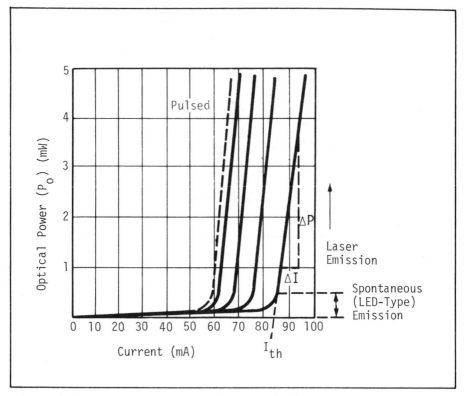

Figure 2.6 - Power vs. Current Characteristics of a Laser Diode at
Different Ambient Temperatures.[14]

2.5 LEDs VS. LASER DIODES

At present, LEDs and injection laser diodes (ILDs) are the most suitable sources for optical-fiber telecommunications. Both are directly modulated by an input current that injects carriers across the p-n junction of a semiconductor diode. However, in designing an optimum system, the special features of each light source should be taken into consideration. In general LEDs and laser diodes, when used in a system application, differ in the following: power levels, temperature sensitivities, response times, lifetimes and characteristics of failure.

2.5.1 Suitability for Fiber Optic System Operation [8,14]

The diode laser is a preferred source for moderate band to wideband (50 to 300-MHz) systems. It offers a fast response time (typically less than 1 ns); has a narrow optical bandwidth (Fig. 2.7), as a rule less than 1 nm; and can couple high levels of useful optical power (usually several mW) into an optical fiber with a small core and a small numerical aperture.

Some systems operate at a slower bit rate and require more modest levels of fiber-coupled optical power (50 to 250 μW). This is the domain of high-radiance LEDs. A LED requires less complex driving circuitry than a diode laser, needs no thermal or optical stabilization and, in general, can be produced more cheaply, with higher yields. In addition, LEDs have longer operating lives and fail in a more gradual and predictable fashion than diode lasers.

Both LEDs and diode lasers behave electrically as diodes, but their light-emission properties differ substantially. Since a small diode laser is an optical oscillator, it has many typical oscillator characteristics, i.e., a threshold of oscillation, narrow emission bandwidth, a temperature coefficient of threshold and frequency, modulation nonlinearities, and regions of instability. These properties, along with the diode laser's maximum power output, efficiency and expected life, affect the design and overall performance of a laser transmitter in an optical communication link.

2.5.2 Comparison of Typical Parameters

Typical parameters for some commercially available LEDs and laser diodes are shown in Table 2.4. With a high differential quantum efficiency in the transition regions, laser emitters can reach several watts of instantaneous power during pulse operation. Even the best LED emitters supply no more than a watt in the pulse mode.

2.13

Table 2.4 - Typical Parameters for Certain LEDs and Laser Diodes (After Ref. 2).

	Light-Emitting Diode			Laser		
				SH-CC*		LOC**
Crystal	GaAs	GaAsP	GaP	GaAs	GaAs	GaAs
Temperature (K)	300	300	300	77	300	300
Emission Wavelength (μm)	0.94	0.66,0.61	0.69,0.55	0.85	0.9	0.9
Spectral Bandwidth (μm)	0.05	0.03	0.09,0.03	0.015	0.015	0.015
Typical Drive Current	<50mA	10-20mA	10-20 mA	<5A	<30A	15A
Mode (pulsed or continuous)	p,cw	p, cw	p, cw	p	p	p
Maximum Pulse Duration	any	any	any	2μs	0.2μs	0.1μs
Duty Factor (%)	any	any	any	<2	<0.1	<1
Rise Time (ns)	<300	10	300	<1	<1	<1
Typical Output (cw)	2.5 mW	1700 nit (500fL)	60-3500nit (175-1000fL)	–	–	–
Peak Pulse Power (W)	–	–	–	5	12	5
Power Efficiency (%)	3	<1	3, 0.1	20-40	3-5	5-10
Beam Spread	Surface or Edge 80°; Dome 180°			15° X 25°		

* SH – Single Heterojunction

   CC – Close Confinement

** LOC – Large Optical Cavity

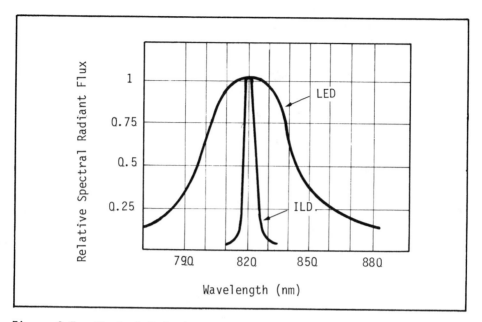

Figure 2.7 - Typical Emission Spectra for Diode Lasers and LEDs Show the Relatively Narrow Bandwidth of an ILD above the Lasing Threshold.

Where dc diode laser's power, speed, or spectral width are not essential, LEDs can offer system designers better temperature stability and reliability at lower cost.

2.5.3 Reliability Considerations for LEDs and ILDs

In any practical optical communications system, component reliability is of great concern. It has been a major research goal to identify and correct the many failure mechanisms that were affecting early electroluminescent devices.

LEDs work in the spontaneous emission mode and at lower current densities than lasers. Therefore, they degrade at a slower rate. The operating lives of LEDs are in the range of 10^6 to 10^7 hours.[15]

Lasers, on the other hand, require high localized power density, which puts great strain on the material, causing heating effects and mass transport by current flow. Therefore, they are less reliable than LEDs. However, outstanding progress in increasing lifetime has been achieved since initial room-temperature, continuous-wave (CW) operation was attained in 1970. Manufacturers now quote a lifetime of at least 10,000 hours. Performance of lasers in experiments is consistent with a mean lifetime of the order of 10^5 hours, and prospects appear favorable for achieving 10^6 hours in the future.[14]

2.6 OPTICAL DETECTORS

A wide variety of solid-state photodetectors are available. These devices may be classified as photoconductive or photovoltaic types. Typical solid-state detectors are p-n junction photocells, p-n-p phototransistors, avalanche photodiodes, p-i-n photodetectors, and Schottky-barrier devices.

2.6.1 Characterization of Optical Detectors

Optical detectors convert incident radiant power into current. This process is characterized by a responsivity or sensitivity function $\rho(\lambda)$ which gives the output current I (in amperes) per unit of incident radiant power, P_r, at a given wavelength,

$$\rho(\lambda) = \frac{I}{P_r(\lambda)} \tag{2.7}$$

For monochromatic radiation, the radiant power in watts is given by:

$$P_r = \frac{I_p hc}{\lambda} \tag{2.8}$$

where,

I_p = photon current ~ photon/sec.

h = Planck's constant = 6.625×10^{-34} joule/sec.

c = velocity of light = $2.998 \times 10^{+8}$ m/sec.

λ = wavelength in meters

At a given wavelength, the sensitivity can be defined in terms of the ratio of output electron current to the number of input photon currents or quantum efficiency, $\eta = I/I_p$. Substituting Eq. (2.8) into Eq. (2.7), gives:

$$\rho(\lambda) = \frac{I\lambda}{I_p hc} = \frac{\eta\lambda}{hc} \tag{2.9}$$

Substituting values for the constants, h and c and expressing λ in micrometers yields:

$$\rho(\lambda) = 0.8 \, \eta\lambda \text{ amp/watt} \tag{2.10}$$

Table 2.5 gives some of the more common detector terms and definitions.

Table 2.5 - Common Detector Terms and Definitions[16]

Term	Symbol	Definition
Sensitivity	$\rho(\lambda)$	The ratio of output current to the incident optical power measured in amperes/watt at a given wavelength.
Noise	V_n	Extraneous voltage generated by a detector.
Johnson or thermal noise	V_J	Noise due to random motion of electrons in a resistance element. The root-mean-square Johnson noise can be approximated as $V_J = \sqrt{4KTR\Delta f}$. Boltzman's constant, K, is 1.38×10^{-23} joules/kelvin, T is the absolute temperature in degrees kelvin, R is the resistance in Ohms, and Δf is the electrical bandwidth of the circuit in Hz.
Noise equivalent power	NEP	The incident radiation in watts required to produce an output signal equal to the detector noise.
Detectivity	D	The reciprocal of NEP .
Specific detectivity	D*	Detectivity normalized for a detector area of $1 cm^2$ and a bandwidth of 1 Hz. It is equal to $D\sqrt{A\Delta f}$, where A is the detector active area in cm^2 and Δf is the bandwidth in hertz. When reporting D*, the wavelength or the black-body temperature and chopping frequency at which the data are taken must be specified.
Time constant		The period required for a detector to reach 63.2% of its final output value following the application of steady-state incident radiation. The time constant equals $1/2\ \pi f$, where f is the chopping frequency in Hz at which the frequency response begins to roll off at 3 dB/octave.

2.6.2 Factors Determining the Response of Optical Detectors [17, 18]

The main factors that determine the response of the detector to incident light are:

- *Thickness of the sample:* For thin samples some photons will get through without generating a pair.

- *Lifetime of carriers:* If a photon-generated minority carrier is far from the junction and its lifetime is low, it may recombine with a majority carrier before reaching the junction.

- *Surface perfection:* A semiconductor surface with its dangling bonds can accumulate charges which cause carriers generated in its vicinity to recombine immediately. The surface lifetime (which is a measure of the surface state density) must be made large if good blue response is required.

- *Passivating oxide thickness (antireflection film):* A thin layer of silicon dioxide, silicon nitride, or aluminum oxide can act as an antireflection oxide to allow most of the light in a given wavelength range to be transmitted into the silicon. The thickness of the oxide will determine which wavelengths are passed and which are reflected.

- *Diffusion depth:* This parameter plays an important role in that it establishes the amount of energy that provides wavelengths which can generate electron-hole pairs.

- *Wavelength of the light:* This physical parameter is triply important because its value influences the following optical parameters: (a) depth of penetration, (b) antireflection and (c) relative response.

2.6.3 PIN Diodes vs. Avalanche Diodes

P-i-n photodiode and *avalanche* photodiode devices both have excellent efficiency for conversion of light energy to electrical responses necessary to handle the bandwidths involved when considering fiber optic communication systems.

A *p-i-n* photodiode, or simply photodiode, has two terminals, a cathode and an anode. It has low forward resistance (anode positive) and high reverse resistance (anode negative). In contrast, a

positive bias is required for active areas of *avalanche* photodetectors.

The simplest model of a photodiode is just an ordinary diode having a current source in parallel as in Fig. 2.8.[3] The magnitude of
this current source is proportional to the radiant flux being detected by the photodiode. The polarity of this photocurrent is from
cathode to anode. Thus, it is apparent that if zero external bias is
applied, the photocurrent will cause the anode to become positive with
respect to the cathode. Part of the photocurrent will flow back
through the photodiode, and part will flow in the load resistance.
If the load resistance is open or extremely high, most of the photocurrent flows in the forward direction through the diode. This may
seem paradoxical, but such a model does describe the results obtained
in an open circuit.

Operation with zero bias is called the photovoltaic mode because
the photodiode is actually generating the load voltage. Photovoltaic
operation can be either logarithmic or linear depending upon the
value of the load resistance. Logarithmic operation is obtained if
the load resistance is very high ($>10^{11}$ ohms). Linear operation is

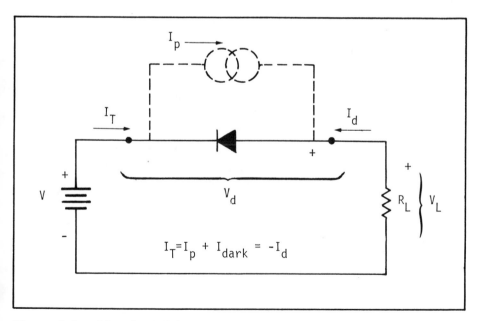

Figure 2.8 - Circuit Model of a PIN Photodiode.

obtained if the load resistance is very low with respect to the dynamic resistance of the photodiode. The upper limit of linear zero-bias operation is at $V_L \approx 100$ mV, depending on the precision of the linearity requirement. With higher values of R_L, sensitivity can be increased for detecting very low-level signals, but the dynamic range of linear response is decreased. The maximum practical value of R_L ranges from 25 MΩ for large area photodiodes to 550 MΩ for the smaller devices.

Operation with reverse bias is called the photocurrent or the photoconductive mode. Practical photoconductive mode circuits are discussed in Chap. 6.

Avalanche photodiodes (APDs) are so called due to the known avalanche effect in which the photocurrent is multiplied. When the device is operated near the reverse breakdown voltage, V_{BR}, the carriers (electrons or holes) that are moved toward the junction are accelerated to such an extent that they ionize the lattice, generating secondary hole-electron pairs, which in turn are accelerated, generating more hole-electron pairs. The result is a photocurrent gain, M, which may be 10 to 1000 in silicon avalanche photodiodes. It is this photocurrent signal gain, occuring before the introduction of the preamplifier noise, that makes the APD ideally suited to high sensitivity applications. In addition to having a high gain factor, APDs operate at a very high frequency, making them ideal detectors for long distance fiber-optics telecommunication systems, which operate with low light levels and wide bandwidths. Other characteristics of APDs include noise, capacitance, output current, response time and temperature coefficient.

Typical device parameters for the two types of optical detectors, considered here, are shown in Table 2.6.

Table 2.6 - PIN vs. APD Typical Parameters[18]

DEVICE PARAMETERS	P-I-N PHOTODIODE	AVALANCHE PHOTODIODE
Active area	0.2 mm^2 to 5 mm^2	-0.5 mm^2
Responsivity	~ 0.5 A/W	30 to 60 A/W
Dark Current (DC)	1 to 10 nA	10 to 100 nA
Reverse bias	≤ 100 V	≤ 400 V
NEP	$\geq 10^{-13}$ W$\sqrt{\text{Hz}}$	$\geq 10^{-15}$ W$\sqrt{\text{Hz}}$
Rise-Time	< 4 ns	< 2 ns

2.7 PIGTAILING ELECTRO-OPTICAL SOURCES AND DETECTORS

In communications fiber optic systems, the designer may consider the case of using housed LEDs or lasers as well as photodetectors, with a fiber *pigtail*, ready for coupling into a fiber-optic link. Some manufacturers have already begun to offer light sources and detectors with *pigtail* fibers attached. The purpose of these devices is to eliminate the problem of coupling a discrete photodetector, LED, or laser diode to a small diameter optical fiber.

With pigtailing, semiconductor sources and detectors can be placed anywhere on a board, without preventing the *pigtail* ends from merging into one common I/O port. In addition, pigtailing provides a low cost connection with repeatable insertion losses.[19] Otherwise, it would be difficult to make a sensible multichannel fiber-optic connector without using *pigtails*. For example, where individual semiconductor elements are integrated with the connector body, it would be difficult to construct a multichannel I/O port without putting together an impossibly massive connector. Figure 2.9 shows some representative methods for pigtailing optoelectronic devices.[20]

Of course, when using these devices, it should first be determined whether the *pigtail* is compatible with the fibers to be used and whether the amount of power in the fiber is adequate to do the job at the receiving end. More about optimization of coupling between a fiber and the source is discussed in Chap. 4.

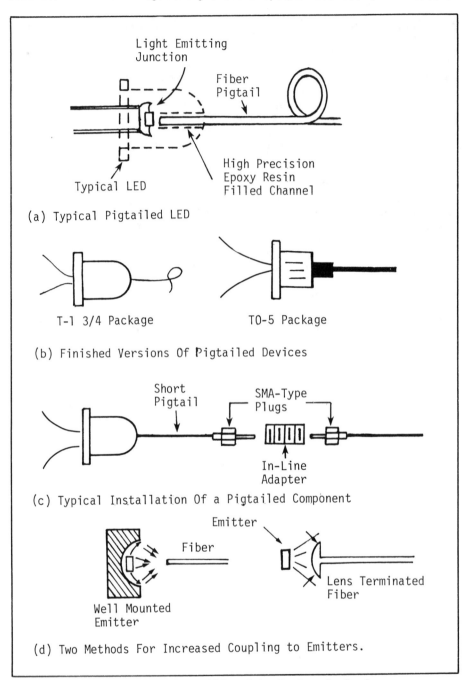

(a) Typical Pigtailed LED

(b) Finished Versions Of Pigtailed Devices

(c) Typical Installation Of a Pigtailed Component

(d) Two Methods For Increased Coupling to Emitters.

Figure 2.9 - Details of Pigtailing Optoelectronic Devices
(After Ref. 20).

2.8 LIGHT COUPLING IN NON-CONTACTING DEVICES AND SPATIAL SYSTEMS

The ability to transmit data through the air (or free space) on an optical beam presents an attractive alternative to stringing cables, either electrical or fiber optics. For very small separations, as in the case of optical couplers, there are no significant problems. As the separation or distance between light source and photodetector becomes larger, the spreading behavior of the light waves becomes a factor with which to cope.

2.8.1 Open and Closed Optical Systems

Optical systems can be either open or closed transmission systems, and have their respective preferred applications and limitations. With an open system, light transmission is based on the principles mentioned above. The small couplers relate to very small source-detector distances, which lie below the critical photometric distance. Those for short, medium or long ranges relate to source-detector distances above the photometric critical distance.

Closed optical systems are optical fibers, or lightguides, that can bend to varying degrees depending on the fiber, without affecting the light path. Their operation is governed by the phenomenon of total internal reflection within the optical element.

Open path systems are discussed in this section. The principles of fiber optic transmission are covered in Chap. 4.

2.8.2 Characteristics of the Transmission Path

A source flux coupling to a receiver is dependent upon the source radiation pattern, the source receiver spacing, and the receiver area. In most system analyses, the available flux and the responsivity of the receiver are considered to be constant for each application. The dynamic incidence change found at the receiver results from modifications of the transmission path. Figure 2.10 shows various techniques of flux path modulation.

Optical power attenuation takes place due to a number of reasons. One is the transmission loss as the flux is incident upon and passes through the transmission media. These losses are due to reflection at the surface of the material, and scattering and absorption within the material. For the calculation of the optical transfer function, these losses are designated by τ and the transmittance of the material for the wavelength of the source by T. The relationship, then, for these quantities is:

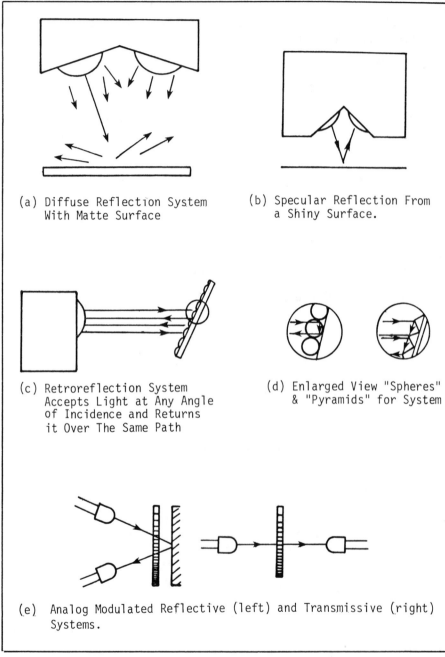

(a) Diffuse Reflection System With Matte Surface

(b) Specular Reflection From a Shiny Surface.

(c) Retroreflection System Accepts Light at Any Angle of Incidence and Returns it Over The Same Path

(d) Enlarged View "Spheres" & "Pyramids" for System

(e) Analog Modulated Reflective (left) and Transmissive (right) Systems.

Figure 2.10 - Diffuse and Specular Reflection Systems in Light Beam Transmission Systems.

$$T = 1 - \tau \qquad (2.11)$$

2.8.3 Optical Transfer Function[21]

The characteristics of the transmission path can be predicted through the use of an optical transfer function, OTF. This transfer function may be defined as the ratio of the total optical power available at the source, ϕ_S, to the incident flux arriving at the receiver ϕ_R, i.e.:

$$OTF = \frac{\phi_R \ (\text{Received})}{\phi_S \ (\text{Sent})} \qquad (2.12)$$

The amount of flux coupled into the receiver area A relates to the relative aperture of the receiver. The relative aperture, also called numerical aperture, N.A. is defined as the sine of one-half the included angle of the receiver cone (Fig. 2.11). Thus,

$$N.A. = \sin\theta \qquad (2.13)$$

where $\theta = 1/2$ cone angle. In practical applications, the use of a lens will provide a more efficient coupling.[21]

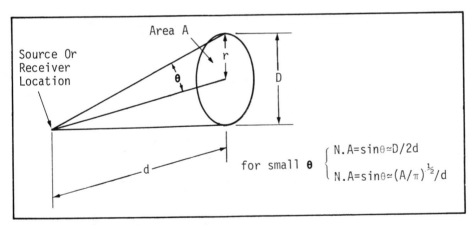

For small θ
$$\begin{cases} N.A. = \sin\theta \approx D/2d \\ N.A. = \sin\theta \approx (A/\pi)^{\frac{1}{2}}/d \end{cases}$$

Figure 2.11 - Definition of Numerical Aperture (N.A).

2.8.4 A System Model for an Optical Coupler

The electrical circuit for an optical isolator is shown in Fig. 2.12a. A packaging sketch is shown in Fig. 2.12b, while a system

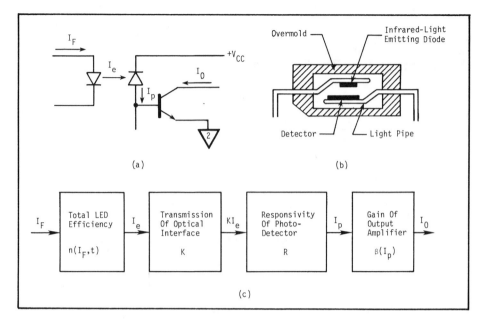

Figure 2.12 - The Electrical Circuit (a), a Packaging Sketch (b) and the System Model (c) for an Optical Isolator.

model for the device is shown in the block diagram of Fig. 2.12c. From this block diagram of the model, the expression for the coupling transfer ratio (CTR) can be written as follows:[22]

$$CTR = \frac{I_O}{I_F} \ (100\%) = KR\eta(I_F,t) \ \beta(I_p,t) \tag{2.14}$$

Optically coupled isolators are useful for applications where analog or dc signals need to be transferred between two isolated systems in the presence of a large potential difference or induced noise. Examples include sensing circuits (thermocouples, transducers, etc.), patient monitoring equipment, power supply feedback, high-voltage current monitoring, and audio or video amplifiers. In many applications, the isolator can transmit the analog signal directly. Due to the wide use of optical couplers in these and other applications, Chap. 3 is devoted to a more thorough discussion of these devices.

2.8.5 Line-of-Sight Optical Communications [23]

There are some differences between electro-optical terminals for line-of-sight and fiber optic systems. The main difference lies in the method of coupling the modulated optical beam into and out of the

transmission medium. A short-haul (less than 1 km) atmospheric system includes a transmitter and a receiver. A block diagram of such
a system is shown in Fig. 2.13. In this direct optical-beam transmission system, a lens is used to capture the emitted radiation (signal) and launch it into the atmosphere. Similarly a lens is used at
the receiver end. The other basic system elements are transmitter/
receiver electronics and LED/PD diodes, as in a fiber optic link system.

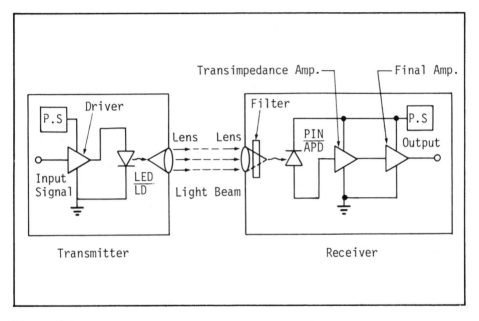

Figure 2.13 - Direct-Optical-Beam Transmission System.

    Typical radiated power for the kind of atmospheric system dis-
cussed in the 0.8 to 0.9 µm wavelength range varies from 1 to 20 mW
for IR-LED devices with 10 ns rise time. Although atmospheric systems suffer from disruptive effects that are not encountered in
fiber optic cable installations (e.g. background noise, fading, and
unpredictable attenuation), satisfactory performance has been reported for existing systems in operation. In space applications, laser
links offering a high degree of directivity and privacy are basically
used for space-space, space-aircraft, and space-ground communications.

2.9 SOLUTIONS TO RADIATION, CROSSTALK AND GROUNDING PROBLEMS VIA LIGHT COUPLING

Among non-contact, non-destructive detectors, optoelectronic devices are superior in that they need not be intimately close to the sensed object, as most inductive, capacitive, RF, ultrasonic and magnetic-pickup-type proximity devices must be. Every optoelectric system referred to in this section consists of a light detector (sensor), an amplifier, and an output. Inherently, optical systems are electrically isolated, meaning they are immune to electromagnetic, RF, and crosstalk interference. In addition, optically coupled devices may be used to protect delicate circuits from accidental overvoltages and eliminate system ground loops.

2.9.1 Spatially Parallel Channels

The relative ease of manipulating optical information in spatially parallel channels becomes attractive when considering linkages over short distances. A simple lens can image an array of light emitters such as a matrix of individual lasers onto another matrix of light detectors (Fig. 2.14a). Such a system is compared with a conventional hard-wire system in Fig. 2.14b. The optical system in Fig. 2.14a solves the problems of radiation, crosstalk and ground loops encountered in the system of Fig. 2.14b, but it becomes impractical when the distance between the emitters and detectors is increased, since the image will be degraded due to diffraction losses and scattering effects.

2.9.2 System of Optical Scanners

Figure 2.15 shows a two-scanner system, where the principles of *retroreflecting* and *beam splitting* are applied to solve the common problems of signal detection. In conventional designs, the output switching transistor is usually connected directly into an internal power supply and is thus not isolated. This characteristic can cause problems when two or more scanners are connected together for AND or OR operation. A remedy for this problem involves using an isolator for each scanner, in this case an optical isolator, as shown in the output of each amplifier.[24]

2.9.3 Coupling or Isolating Functions

Most of the current and prospective application areas utilize two separate capabilities of optoelectronic devices: coupling and isolating. Whereas optocouplers simply form an interface between

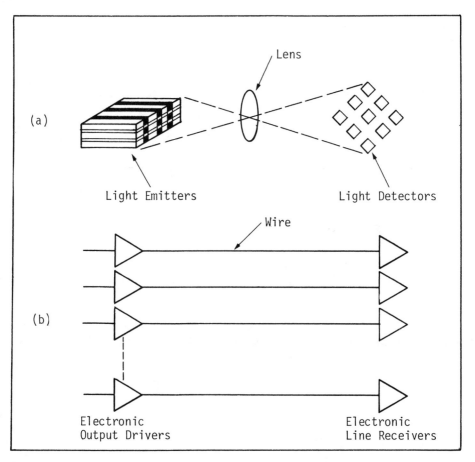

Figure 2.14 - Comparison of Data Transmission Using Parallel Channels: a) by Optical Interconnection Between Light Emitters and Light Detectors and b) by Hard-wire Interconnections Between Electronic Circuits.

two electrically independent circuits (configured from two incompatible logic families such as TTL and CMOS, for example), optoisolators must protect a circuit from potentially damaging voltage transients.

In reality, the only difference between optoisolators and optocouplers is one provided by marketing people. Basically, these optoelectronic components change names to suit the particular market segment in question.[25]

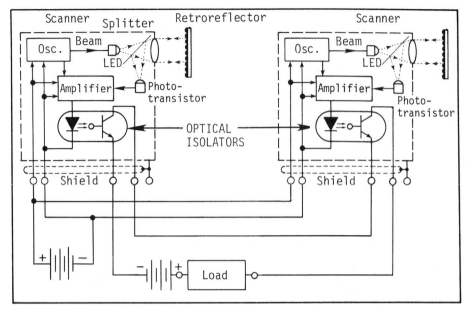

Figure 2.15 - Optical Isolated Outputs for Each Scanner Provide   the
Required Isolation in this System (Adapted from Ref. 24).

2.10 LIMITATIONS OF OPTICAL COUPLING DEVICES

Optical-power and bandwidth limitations, which are of significance in an optical-fiber system, as will be seen in other chapters
of this book, also depend on the light source (modulation rates, radiance, optical-coupling efficiency) and detector (responsivity, rise
time, noise factors) characteristics. Some of the limitations are
discussed below for each device or system separately.

2.10.1 Limitations of Light Sources

Common LEDs possess linewidths of 200-300 $\overset{o}{A}$ which is about one
order of magnitude broader than a multimode diode laser. Because of
the spectral dispersion in fibers, this broad spectral width limits
the information capacity that can be launched using LEDs in long
distance communications. To solve this problem, considerable effort
has been devoted to the development of high-brightness LEDs such as
the successful *Burrus-type surface emitters* and *edge-emitter LEDs*.[8,9]

There are various limitations[15] which must be taken into account
when considering the means of modulation for injection diode lasers.
The first limitation is *thermals*.[26] There is a practical limit to
the power which can be dissipated by the laser chip via its heat sink.
Moreover, the threshold current increases with junction temperature,
thus requiring more drive current which increases the junction temperature. Therefore, special circuitry is required for thermal stabilization.

The second limitation is that of *peak optical power output*. The
laser output may become *noisy* at high power levels and these regions
should be avoided if system performance is not to be impaired.

The third limitation is that of *turn-on delay*. If a pulse step
of current is applied to a laser to take it above threshold, there
may be a delay of a few nanoseconds before laser emission appears.

2.10.2 Limitations of Photodetectors[17,27]

In the receiver circuit, the signal current must contend with
noise currents. In most *PIN photodetectors*, the dominant noise
component is caused by fluctuations in *dark current* - the current
that flows through the diode-biasing circuit when no light is incident
on the photodiode. Dark current increases with temperature and, as
a general rule, it doubles for every 10°C increase in operating
temperature. *Noise equivalent power (NEP)* is another photodiode figure of merit related to noise performance and is expressed in $W/\sqrt{Hz}$.

Multiplying NEP by the square root of the detector-noise bandwidth B, an absolute power called *minimum detectable signal (MDS)* is obtained. The MDS defines the optical power incident on the photodiode which is required to generate a photocurrent equal to the total photodiode noise current (i.e., a 0 dB signal-to-noise ratio). To ensure signal integrity, the receiver should be operated at a level higher than the MDS. Additionally, because the current generated by the photodiode is so small, a pre-amplifier (transimpedance or op-amp) is also a noise source not included in the photodiode's NEP.

Avalanche Photodiodes (APDs) which, like PIN photodiodes are using reverse bias in receiver circuits, but require a considerably higher voltage than PIN diodes. Reverse voltages on the order of 300 volts are not uncommon for the APD's. Due to this high voltage, the average gain fluctuates and creates what is called *excess noise* - a factor limiting the APD's usable gain. In addition, *the temperature dependence* of gain, and hence responsivity, is an operational problem with APDs. Therefore, for general-purpose use in a receiver, an APD requires some form of *automatic gain control.*

2.10.3 Limitations in Linear Operation of Optical Couplers

A single LED and photodiode combination, while useful in a wide range of digital isolation, switching and counting applications, has fundamental limitations that make it impractical for most linear applications - primarily due to nonlinear characteristics and instability as a function of time and temperature.

Nonlinearities are inherent in the basic current-to-light and light-to-current relationships. Gain accuracy is affected by both temperature change and decreasing LED light output with age. Although compensation circuitry can be designed to overcome these effects, circuit costs usually outweigh the advantages of optical coupling. One approach that solves these previous limitations, in a relatively simple and inexpensive way, is to use two matched photodiodes.[28]

2.10.4 Coupling Capacitance in Input-Output Signal Lines of an Optical Isolator [3]

Isolation in optical isolators is affected by stray capacitive coupling as shown in Fig. 2.16. Capacitance C_{CM} is only a small part of the total input-to-output capacitance C_c, which usually is given in data sheets as C_{I-O}. C_c is also called reverse coupling capacitance. It should be noted that C_{CM} is fairly constant and in most optical isolators is between 0.05 and 0.1 pf, whereas in the internally shielded types it is 10 times smaller.

16. Molle, C.S., et al; *Use Low-Cost IR Detectors;* Electronic Design 12, pp. 84, June 7, 1975.

17. Shah, A.; *Detectors on a Parameter Ballot;* Optical Spectra, Nov. 1975.

18. Pitcher, J.; *Considerations of Optical Fiber Measurements;* Electronic Engineering, p. 51, Mid-April 1980.

19. Ohr, S.; *Fiber-Optic Semis Carve Out Wider Infrared Territory;* Electronic Design 2, p. 62, January 18, 1980.

20. Math, I.; *Pigtailing Electro-Optical Sources and Detectors;* Electro-Optical Systems Design, p. 36, September 1978.

21. Hewlett Packard Optoelectronics Division; Optoelectronics Applications Manual Supplement, pp. 15.3-15.6, Palo Alto, Cal.

22. Hewlett Packard Optoelectronics Division; *Consideration of CRT Variations in Optically Coupled Isolator Circuit Design;* Optoelectronics Designer's Catalog 1980, p. 415, Palo Alto, Ca., 1980.

23. Melved, D.B.; *Line-of-Sight Optical Communications;* Electro-Optical Systems Design, Vol. 10, No. 11, pp. 78-81, 1978.

24. Filichowski, W.; *LED Photoelectrics;* Digital Design, pp. 40, March 1977.

25. Clarke, G. J.; *Data Protected Over Long Line Transmission;* Control and Instrumentation, p. 7, April 1979.

26. Laff, R.A., et al; *Thermal Performance and Limitations of Silicon - Substrate Packaged GaAs Laser Arrays*; Applied Optics, Vol. 17, pp. 778-784, March 1, 1978.

27. Zucker, J.; *Choose Detectors for Their Differences, to Suit Different Fiber-Optic Systems;* Electronic Design 9, pp. 165-169, April 26, 1980.

28. Burr-Brown; *Breakthrough in Optical Coupling Results in Small, Low-Cost Isolation Amplifier;* Update, Vol. II, No. 2, pp. 1-2, June 1976.

CHAPTER 3
OPTICAL ISOLATORS

Perhaps the most versatile of any light-activated component, optical isolators (also called optocouplers), play a major role as isolation elements in control systems, in telephone communications and as line receivers in digital data communications. Moreover, the components have many linear applications - for example, as differential isolation amplifiers, ac-coupled amplifiers and servo-isolation amplifiers.

3.1 INTRODUCTION

An *optical isolator* consists of a photon emitting device and a photo-sensitive detector. In the optical isolator, or photon coupled pair, the coupling is achieved by light being generated on one side of a transparent insulating gap and being detected on the other side of the gap without an electrical connection between the two sides (except for a coupling capacitance of approximately 1 pF). In a typical optical isolator, the light is generated by an infrared light emitting diode (LED) and the photo-detector is a silicon diode, transistor, SCR or Darlington, as shown in Fig. 3.1.

Optical isolators have a host of applications. Following is a list of some of their uses.

- Isolate different voltage levels between circuits.

- Prevent interference between control and power circuits, using the unidirectional feature.

- Insulate people or low-voltage circuits from the hazards of high voltage shock.

- Eliminate dc ground loops.

- Amplify or attenuate signals.

- Perform on/off switching.

3.1

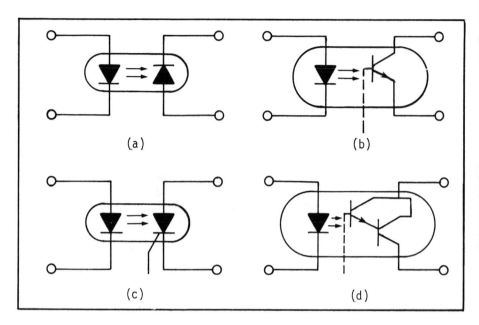

Figure 3.1 - Basic Types of Optical Isolators, a) LED-Photodiode, b) Phototransistor with or without Base Terminal, c) LED-Photo-SCR and d) LED-Photo-Darlington.

3.2 DEFINITIONS AND THEORY

Optical isolators are an extension of the simple optical switch. A light source and a detector are placed in a common housing, and light from the source (usually an infrared LED) causes a flow of current in the detector. The versatility of the device stems from the variety of semiconductor circuits and devices which can serve as a detector.

The sensitivity of the silicon material peaks at the wavelength emitted by the LED, giving maximum signal coupling. The light source and photodetector must have matching spectral characteristics to optimize the operation of an optical switch or isolator. As shown in Fig. 3.2, the spectral response curves of two opto-devices, one photodetector and one LED source, intersect at a wavelength of 900 nm.[1]

Consider a typical optical isolator consisting of a Gallium Arsenide infrared emitting diode, and a silicon phototransistor mounted together in a package, as for example in a DIP (Fig. 3.3).

When forward current (I_F) is passed through the Gallium Arsenide diode, it emits infrared radiation peaking at about 900 nm wavelength. This radiant energy is transmitted through an optical coupling medium and falls on the surface of the NPN phototransistor.

Phototransistors are designed to have large base areas and hence a large base-collector junction area and a small emitter area. Some fraction of the photons that strike the base area cause the formation of electron-hole pairs in the base region. This fraction is called the *Quantum Efficiency* of the photodetector.

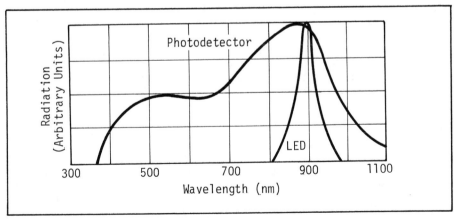

Figure 3.2 - LED/Detector Matching Characteristics.

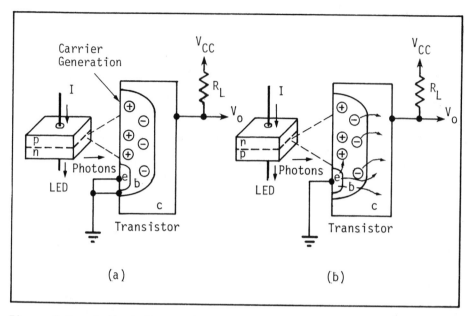

Figure 3.3 - Radiant Energy from a GaAs Infrared LED Diode Activates a Phototransistor, a) With Grounded Emitter and Base, b) With the Base Connection Left Open.

If the base and emitter are grounded (Fig. 3.3a) and a positive voltage is applied to the collector of the photo-transistor, the device operates as a photo-diode. The high field across the collector base junction quickly draws the electrons across into the collector region. The holes drift towards the base terminal attracting electrons from the terminal. Thus, a current flows from collector to base, causing a voltage drop across the load resistance, R_L.

The high junction capacitance, C_{cb}, results in an output circuit time constant $R_L C_{cb}$, with a corresponding output voltage rise time. The output current in this configuration is quite small and hence this connection is not normally used.

The most common circuit configuration is to leave the base connection open. With this connection (Fig. 3.3b) the holes generated in the base region cause the base potential to rise, foward-biasing the base-emitter junction. Electrons are then injected into the base from the emitter to try to neutralize the excess holes. Because of the close proximity of the collector junction, the probability of an electron recombining with a hole is small and most of the injected electrons are immediately swept into the collector region. As a result, the total collector current is much higher than the photo-

3.4

generated current, and is in fact β* times as great.[2] The total collector current is then several hundred times greater than for the
previous connection.

This gain in collector current, however, carries with it a pen-
alty of much slower operation. Any drop in collector voltage is coupled to the base via the collector-base capacitance in the original
photocurrent. Thus, the rate of change of the output voltage is the
same for both the diode and transistor connections. In the latter
case, the voltage swing is β times as great; therefore, the total
rise time is β times as great as for the diode connection. Thus,
the effective output time is β $R_L C_{cb}$.

---

* β is the current gain of a transistor connected as a grounded-
emitter amplifier.

3.3 CHARACTERISTICS OF OPTICAL ISOLATORS

Prior to discussing various applications of optical isolators, it is necessary to review their main characteristics, their driving requirements and their switching responses.

3.3.1 Light Source Input/Output Characteristics

Since the input to all the optical isolators is an LED, the input characteristics will be the same independent of the type of detector employed. Typical LED characteristics are shown in Fig. 3.4. The forward bias current threshold is shown at approximately 1 volt. The current increases exponentially, the useful range of I_F between 1 mA and 100 mA being delivered at a V_F between 1.2 and 1.3 volts. The dynamic values of the forward bias impedance are current dependent and are shown on the insert-graph for R_{DF} and ΔR as defined in the figure. Reverse leakage is in the nanoampere range before avalanche breakdown.

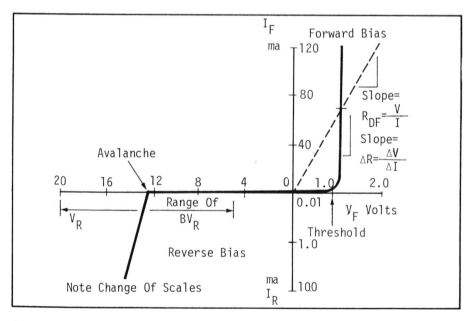

Figure 3.4 - Characteristics of a Typical IR LED.

3.3.2 Driving of Optical Isolator Input Circuit [3,4]

The LED equivalent circuit is represented in Fig. 3.5a where D is an ideal diode. The LED is used in the forward mode. Since the current increases very rapidly above threshold, the device should always be driven in a current mode, not voltage driven. The simplest method of achieving the current drive is to provide a series current-limiting resistor, such as V_B, which is dropped across the resistor at the desired I_F, determined from other criteria. A silicon diode is shown installed inversely parallel to the LED (Fig. 3.5b). This diode is used to protect the reverse breakdown of the LED and is the simplest method of achieving this protection. The LED must be protected from excessive power dissipation in the reverse avalanche region. A small amount of reverse current will not harm the LED, but it must be guarded against unexpected current surges.

When using an optical isolator in a logic-to-logic coupling, a simple transistor drive circuit can be used as shown in Fig. 3.6. In the normally-off case (Fig. 3.6a), the LED is conducting only when the transistor is in saturation. The main design equation for calculating the series current resistor is:

$$R = \frac{V_{CC} - V_F - V_{SAT}}{I_F} \qquad (3.1)$$

The leak current, when the transistor is off, can be bypassed around the LED by the addition of another resistor, R_1, provided its value is such that the leakage current develops a voltage less than threshold V_F. In the normally-on situation (Fig. 3.6b), the transistor is required to shunt the I_F around the LED with a V_{SAT} of less than threshold V_F. The design equation for determining the value of

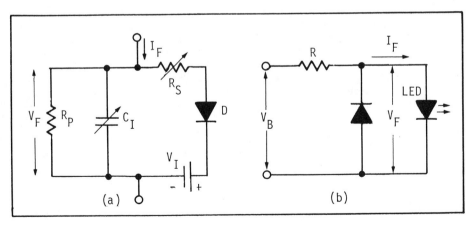

Figure 3.5 - a) LED Equivalent Circuit. b) Typical LED Drive Circuit.

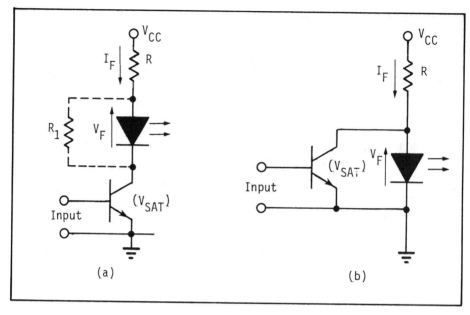

Figure 3.6 - LED Driving Circuits in Logic-to-Logic Coupling, a) Transistor, Normally-Off, b) Transistor, Normally-On.

the series resistor, in this case, is:

$$R = \frac{V_{CC} - V_F}{I_F} \tag{3.2}$$

If the logic is not capable of handling the necessary current, I_F, an auxiliary drive transistor may be used to increase the current capability. Such a representative circuit is shown in Fig. 3.7. Here, a PNP transistor is used as an emitter follower or common collector, to provide a current gain. With the output of gate G_1 low, Q is conducting and current flows through the LED. The design equation for the calculation of R_1 must now include the base-emitter forward biased voltage drop, V_{BE} and the saturation voltage of the gate. Hence:

$$R_1 = \frac{V_{CC} - V_F - V_{BE} - V_{CE(SAT)Gate}}{I_F} \tag{3.3}$$

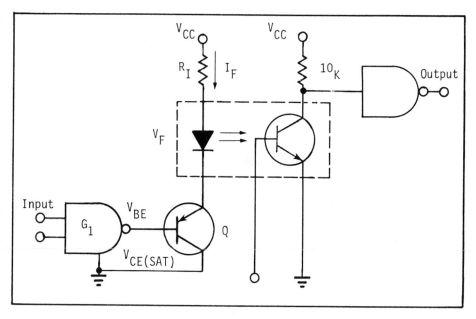

Figure 3.7 - Optical Isolator Driving Circuit with Increased Current Sinking Capability.

3.3.3 Switching Response and Transient Problems

Low-cost optical isolators (Fig. 3.8a) have relatively slow switching speeds, especially for wide voltage swings, but their high speed counterparts are often very expensive. It is possible to increase the switching speed of a low-cost optical isolator by putting in a circuit that provides positive feedback to the light-sensitive base of the device's phototransistor.

For the small price of one transistor, five resistors and a capacitor, it is possible to modify the basic low-cost circuit (see Fig. 3.8b)[5] and improve its response from 2 to 3 kHz to at least 30-kHz for a square-wave output. The collector of transistor Q swings to almost the full supply voltage, with rise times faster than 200 ns and fall times less than 2 μs. Current-switching thresholds for the light-emitting diode remain relatively constant, and there is the added advantage of Schmitt-trigger action in response to slow ramp functions.

Capacitor C_1 is required for correct ac loop gain, with a value of 180 μF giving the proper compensation for best switching speed.

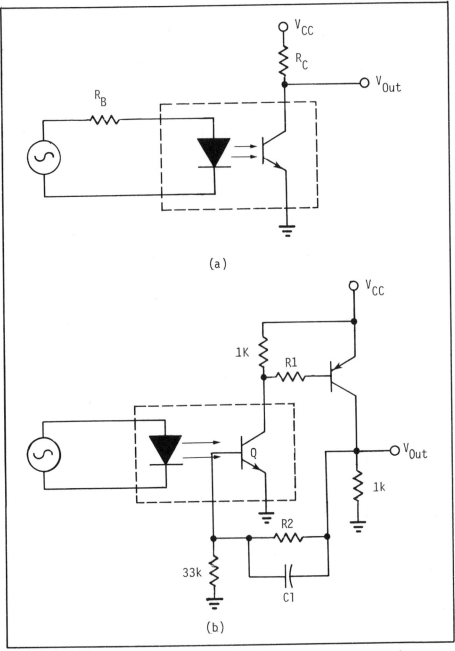

(a)

(b)

Figure 3.8 - a) A Conventional Low-Speed Optical Isolator, b) Modified Circuit Provides High-Speed Operation due to Positive Feedback.[5]

The value of R_2 is directly related to power-supply voltage, and ranges from about 390 kΩ for a 5V supply to 680 kΩ at 12V. At a given supply voltage, small variations in R_2 shift the current-threshold levels (vary the circuit's hysteresis) to compensate for changes in devices.

Many optical isolators have loose current-transfer ratio specs, and must be designed with current overdrive to compensate for device variability. The advantage of this circuit, however, is that it minimizes device-dependent performance while increasing speed.

Insensitivity to industrial noise is an important characteristic in I/O modules.[6] Most modules use optical coupling to avoid noise-induced false triggering. These couplers typically consist of an LED input and phototransistor output. Isolation between input and output is generally good because the parasitic coupling capacitance between the LED and transistor is about 1 pF (Fig. 3.9).

Several module makers have found that optical coupling alone does not provide sufficient isolation from transients. Since the LED and phototransistor are electrically isolated from each other, they can be at different voltage potentials. This potential difference can cause a noise current through the parasitic 1 pF coupling capacitance, C_c. The noise current (which becomes a base current into the phototransistor) is given by I = C(dv/dt).

If the phototransistor has a gain of 100 (a common value for optoisolators), a base current of 0.01 mA will saturate the transistor for the typical configuration shown. From the capacitor current equation, 1 pF and 0.01 mA yields a dv/dt of 10 V/μsec; thus a voltage spike of 10 V/μsec turns on the transistor. Simple relay closures can cause voltage spikes on the order of 10 V/μsec. Semiconductor switching devices such as heaters and switching regulators can cause upwards of 800 V/μsec spikes.

Some modules have a parallel RC network connected between the phototransistor base and emitter.[7] The capacitor provides a path for the transient current, while the resistor provides a discharge path. A typical capacitor value could be 0.01 μF.

3.3.4 Temperature Characteristics and Threshold Adjustment

The forward voltage of the LED has a negative temperature coefficient of 1.05 mV/°C and the variation is shown in Fig. 3.10.

The brightness of the IR LED slowly decreases in an exponential fashion as a function of forward current (I_F) and time.[3] The amount

3.11

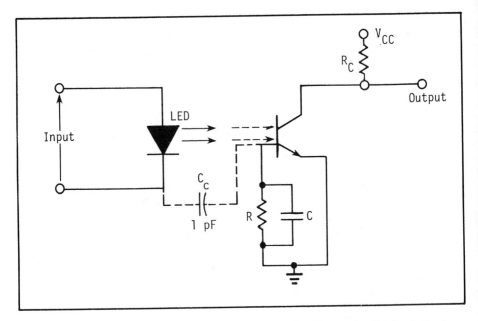

Figure 3.9 - Parasitic Capacitance Creates Problems in Optical Isolators.

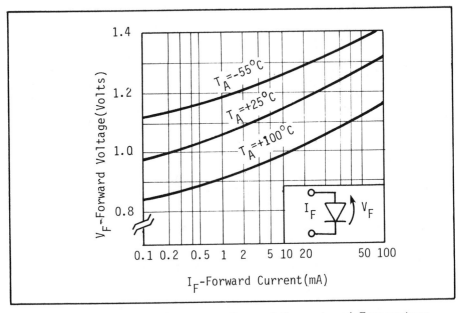

Figure 3.10 - Forward Voltage vs. Forward Current and Temperature.

of light degradation is shown by the graph in Fig. 3.11. This degradation must be considered in the initial design of optical isolator circuits to allow for the decrease and still remain within design specifications on current-transfer-ratio (CTR) over the design lifetime of the equipment. Also, a limitation on I_F drive is shown to extend the useful lifetime of the device.

In some circumstances it is desirable to have a definite threshold for the LED above the nominal V_F diode voltage. This threshold adjustment can be obtained by shunting the LED by a resistor (Fig. 3.12), the value of which is determined by a ratio between the applied voltage, the series resistor, and the desired threshold. The relationships between these values are:

$$I_{FT} = \frac{V_F}{R_2} \text{ and } R_1 = \frac{V_A - V_F}{I_{FT}} \tag{3.4}$$

The calculations will determine the resistor values required for a given I_{FT} and V_A. It is also quite proper to connect several LEDs in series to share the same I_F. The V_F of the series is the sum of

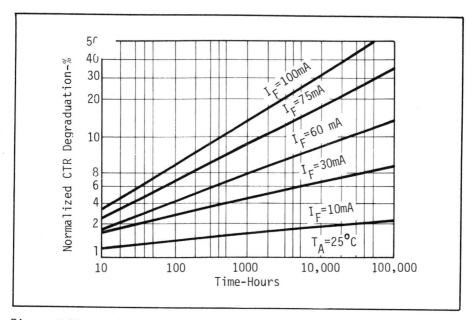

Figure 3.11 - Brightness Degradation vs. Forward Current and Time.

3.13

the individual V_F's.

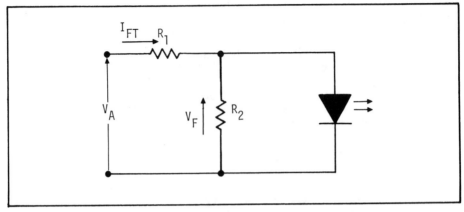

Figure 3.12 - LED Threshold Adjustment

3.4 OPTICAL ISOLATORS IN CIRCUITS AND SYSTEMS

In this section various circuits and systems are presented, where optical isolators (couplers) are used both in digital and analog applications.

3.4.1 Solid-State Relay Circuits

Photodiode coupled pairs (Fig. 3.13) eliminate many undesirable effects in relay circuits such as ground current and spikes. In addition, optical coupling provides 100 GΩ isolation between the TTL and the relay circuit, thus eliminating relay noise and spikes.

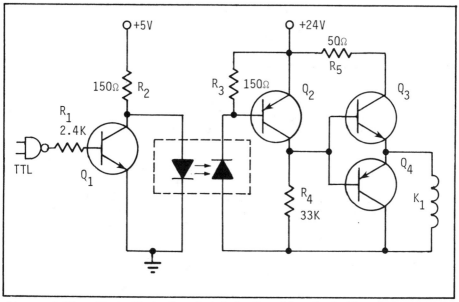

Figure 3.13 - Photodiode Coupled Pair Provides Isolation Between a TTL Circuit and a Relay Circuit

The coupled pair light-emitting diode portion is driven by an input signal which produces light intensity proportional to the signal current. Any input signal variation produces a proportional change at the photodiode output which is normally operated with reverse bias to provide the required voltage swing.

When the TTL output signal is high, Q_1 is turned on, and no current flows in the light-emitting diode; no light to the detector causes the photodiode to be at its 5 GΩ maximum impedance. Therefore, Q_2 cannot conduct, which in turn prohibits Q_3 from conducting. Since Q_3 provides the current path for K_1, the relay will not be energized. When the TTL output is low, the circuit will respond very quickly and the relay energized in a time virtually dependent on its mechanical response alone.

Reed relays have the disadvantage that if they are stuck closed, severe problems in a circuit or system may be created. Most mechanical relays are sensitive to vibration or acceleration forces, when closed as well as when open. Semiconductor relays are not affected by vibration or accelerating forces. The increasing practice of replacing electro-mechanical switching devices with completely electronic solid state switches has developed a need for a small device equivalent to the ubiquitous reed relay.[3,8]

A proper semiconductor relay design must meet isolation, closed circuit current conduction, open circuit terminal voltage, and input drive power specifications. The best isolation is achieved with an optically coupled device, where input-to-output isolation voltages of 1000 to 2500 volts are readily obtained. An input-to-output coupling capacitance of less than 1 picofarad is an added feature.

Typical devices can conduct 100 mA in saturation with only 10-20 mA of input drive and 0.8 volts offset contact voltage. With a current-transfer-ratio (CTR) of 200% to 1000%, these devices are very sensitive as signal amplifiers, and function well as *off-on* switches.

The photo-Darlington opto-isolator circuit is shown in Fig. 3.14. To understand the operation of the photo-Darlington opto-isolator consider the equivalent circuit of Fig. 3.15.[9] The LED is shown inverted to properly relate capacitance connections. The photo-Darlington is an integrated device and both transistors are illuminated by the LED. Hence, both collector-base junctions operate as photo-diode detectors. The isolation coupling capacitance is shown distributed as a result of the internal physical construction of the package lead frame and semiconductor device construction. While the total coupling capacitance is less than 1 picofarad, the high gain of the device tends to amplify the coupling effect to the base node.

Simple relay circuits are represented by Figs. 3.16 and 3.17. In Fig. 3.16 the controlling input transistor conducts I_F in saturation ($V_{CE} \simeq 0.3V$) and I_F is adjusted to meet load current requirements by the value of R_1. For a load current of 150 mA the LED current must be 30 mA or greater. Saturation load curves are provided on the data sheets. In Fig. 3.17 the controlling input transistor also conducts I_F in saturation, thus shunting the current around the LED, which has a threshold voltage over 1.0 volt. Other input control circuits are

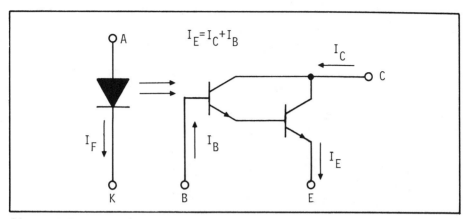

Figure 3.14 - Typical Photo-Darlington Optical Isolator.

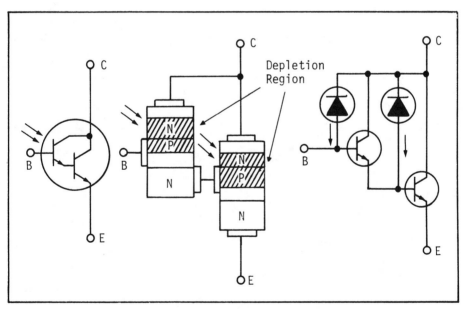

Figure 3.15 - Equivalent Circuit Explaining the Operation of a Photo-Darlington.

possible, and can even be logic gates. A *Form C* (NO/NC) switch can be organized as in Fig. 3.18, where the proper output current polarities must be observed. The saturation voltage of the input transistor

3.17

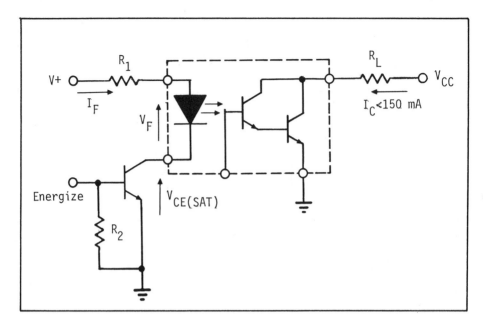

Figure 3.16 - Normally Open Relay Circuit.

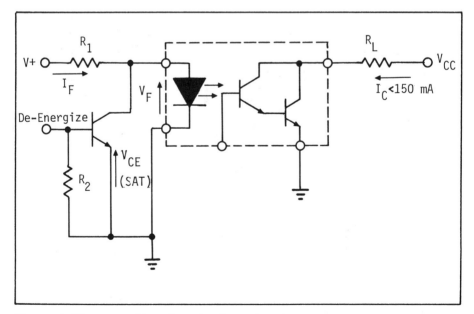

Figure 3.17 - Normally Closed Relay Circuit.

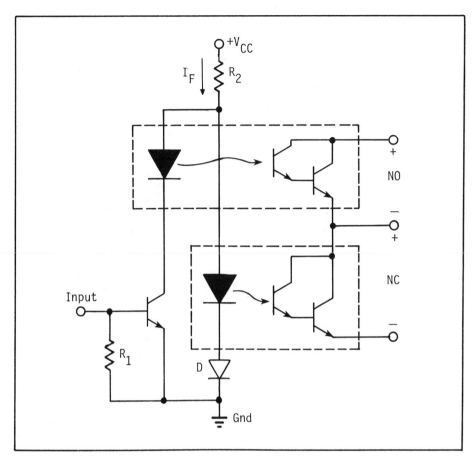

Figure 3.18 - "Form C" Contact Using Two Optical Isolators With Photo-Darlingtons.

is less than the diode (D) voltage drop, effecting commutation .

An extension of these circuit arrangments can be the double-pole, double-throw (DPDT) relay shown in Fig. 3.19. Obviously, relays having many poles may be organized by series connection of the LED inputs. Other circuit configurations are also possible with various desirable features.

It is possible to control a magnetically latched relay with the circuit of Fig. 3.20 that incorporates a dual photocoupled silicon-controlled rectifier (SCR),[10] and eliminates the need for both negative and positive supplies, as well as switching transistors. The key to the circuit is the photo-SCR coupler, which, in effect,

3.19

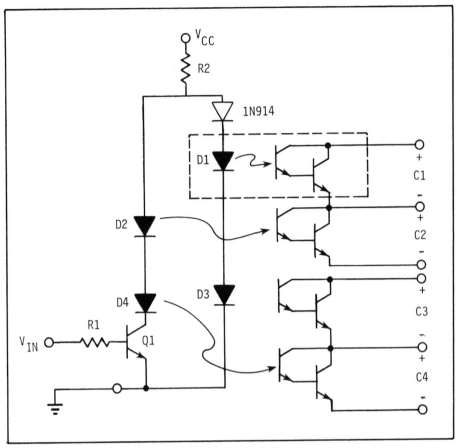

Figure 3.10 - Double Pole, Double Relay Circuit

permits TTL inputs to drive the relay directly. The circuit can actuate a coaxial magnetic-latching relay and latch it in one state with
a 30-V dc input and in the other with the opposite polarity. Switching time from one state to the other is of the order of 30 ms.

The TTL control of the dual photocoupled SCR provides the double-
latching switching function. It may be necessary to add one ac-sup-
ply transformer to the circuit. A varistor across the relay coil
must also be added to suppress inductive transients.

The dual-polarity voltage is obtained by operating the relay
coil on half-wave rectified dc. The SCR turned on by the TTL input
selects the positive or negative half-wave. Since the relay coil
gets only half-wave voltage, the ac transformer voltage selected is

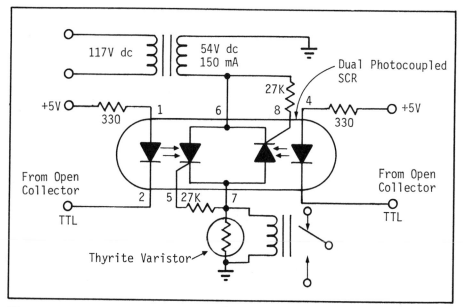

Figure 3.20 - A Dual Photocoupled SCR Passes Current on the Positive or Negative Half-Cycles of Voltage.[10]

somewhat higher, in this case, 54V rms at 150 mA.

In this circuit, the relay operates with 50-ms *ON* pulses. However, the pulse duration can be reduced, because the relay changes state reliably with two half-cycles of 60 Hz or in 16.66 ms.

3.4.2 Power Control Circuits [11]

Control of power circuits with logic requires isolated devices both to provide logic-level inputs from power and mechanical sources and to control power sources from logic output. Figure 3.21 shows TTL-compatible couplers used to interface logic with power circuits. The system monitors both power and mechanical inputs and the common-logic output controls power circuits.

In power-to-logic coupler applications the effects of the huge transients on power lines, which feed through the isolation capacitance, must also be considered. For the SCR coupler (Fig. 3.21a) false triggering could result from these transients, since they have a high dV/dt that can couple into the input circuit. Therefore, a snubber circuit, R_3C_2 must be connected in parallel with the SCR.

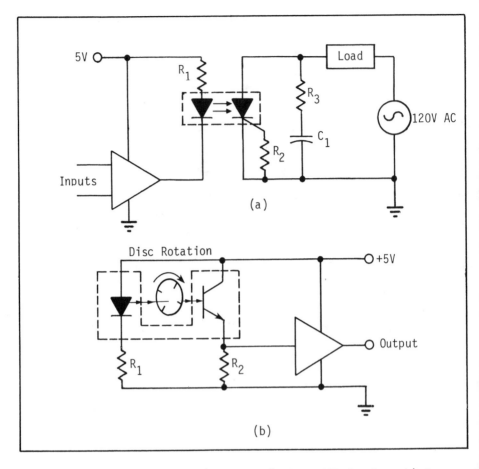

Figure 3.21 - Power Control Circuits, a) Photo-SCR Coupler, b) Interrupter Coupler.

The interrupter coupler for sensing mechanical motion (Fig. 3.21b) is affected by two external sources.

3.4.3 Optical Isolators in Linear Coupling

Optically coupled isolators have proven very effective for eliminating noise and breaking unwanted ground loops in digital circuits. It is perhaps less well known that optical isolators are also suitable for analog circuits.[12] With the correct circuit techniques, the isolator's advantages can be applied to such linear tasks as sensing circuits, patient-monitoring equipment, adaptive controls, power supplies, and audio or video amplifiers.[13,14]

The nonlinearity of the optical isolators (optocouplers) can be eliminated by using coupler pairs with good tracking in their transfer function. A circuit for linear signal transmission consisting of a pair of optical isolators in a feedback arrangement is shown in Fig. 3.22. It consists of a matched pair of LED-photodiode optical isolators operating in the feedback path of an operational amplifier to provide linear signal transmission with stable gain and high-voltage protection. The advantage of this circuit is that the operational amplifier is also protected against high voltage. Hence, a maximum number of components of the circuit are optically isolated from the input port.

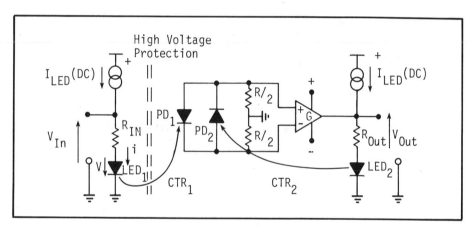

Figure 3.22 - Linear Optocoupler Circuit (CTR = Current-Transfer-Ratio and G = Gain).[13]

Considering a certain range of input current, the transfer function can be approximated as follows;[13]

$$I_{PD} = C(1 + m)^n \qquad (3.5)$$

where, $C = I_{LED}(dc) \cdot CTR$

dc current-transfer ratio $CTR = I_{PD}(dc)/I_{LED}(dc)$

modulation index $m = I_{LED_p}(ac)/I_{LED}(dc)$

Using Eq. (3.5), an expression can be written for the output voltage of the circuit of Fig. 3.22:

$$V_{OUT}(t) = V_{OUT_{dc}}\left\{C_1[1+m_1(t)]^{n_1} - C_2[1+m_2(t)]^{n_2}\right\} \cdot R \cdot G \qquad (3.6)$$

where,

$$V_{OUT}(t) = V_{OUT_{dc}} + v_{OUT} \text{ and } m_1(t), m_2(t) \text{ are input, output}$$

modulation indices, respectively.

With $G \rightarrow \infty$ and $V_{IN}(t) = V_{IN_{dc}} + v_{IN}$, Eq. (3.6) becomes:

$$V_{OUT}(t) = V_{OUT_{dc}} \cdot \left(\frac{C_1}{C_2}\right)^{1/n_2} \cdot \left\{\frac{V_{IN}(t)}{V_{IN_{dc}}}\right\}^{n_1/n_2} \qquad (3.7)$$

When a wideband input, but only moderate accuracy is needed, an analog divider can be used that is simpler and less expensive than a modular or integrated circuit divider. Such a circuit is shown in Fig. 3.23.[15] This circuit is unusual in that it has a single center-tapped photoconductor instead of the two isolated photoconductive elements normally required.

The upper part of the circuit, composed of op amp A_1, R_1 and R_2, is a variable-gain amplifier, with bandwidth limited primarily by the op amp. Its gain, always above unity, is controlled by the voltage divider formed by R_1 and R_2. The resistance of photoconductor R_2 is varied by the LED, which is controlled by a feedback loop that includes op amp A_2, R_3 and R_4.

A photoconductor, R_3, is included in the loop to correct for the nonlinear and poorly defined relationship between the LED current and R_2's resistance. Gain of the upper amplifier (X input) is given by:

$$\frac{V_{OUT}}{X} = \frac{R_1 + R_2}{R_2} \qquad (3.8)$$

Op amp A_2 adjusts the LED current so that:

$$\frac{Y}{V_{ref}} = \frac{R_3}{R_3 + R_4} \qquad (3.9)$$

If $R_2 = R_3$ and $R_1 = R_4$, then:

$$V_{OUT} = \frac{X}{Y} V_{ref} \qquad (3.10)$$

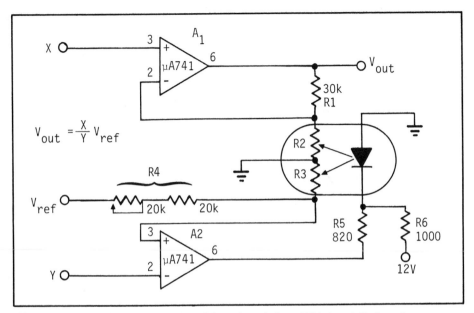

Figure 3.23 - Analog Divider Circuit with a Wideband X Input.

The trimmer part of R_4 corrects for resistor tolerances and photoconductor mismatch. (Tracking accuracy is quoted by one manufacturer as typically 5 to 10%). This trimmer is set by applying a voltage approximately half of V_{ref} to both the X and Y inputs and adjusting for an output equal to V_{ref}. This leaves the errors due to offsets in the op amp and to imperfect tracking in the photoconductors uncompensated. In many cases, the offsets of the op amps will be small and can be neglected.

Originally designed for automatic contrast control in a scanning electron microscope, the circuit should prove useful in acoustically coupled modems, facsimile and audio level control, and other applications requiring an AGC loop. Normal operation requires that the X input be restricted to a value that will not saturate op amp A_1, and that the Y input be kept between zero and V_{ref}. If the Y input goes out of bounds, however, the path between the X input and the output remains linear.

3.4.4 Logic Isolation and Line Receivers

By using an optical isolator between two systems coupled by a transmission line, effective line isolation can be achieved. Figure 3.24 shows a typical interface system using TTL integrated circuitry

3.25

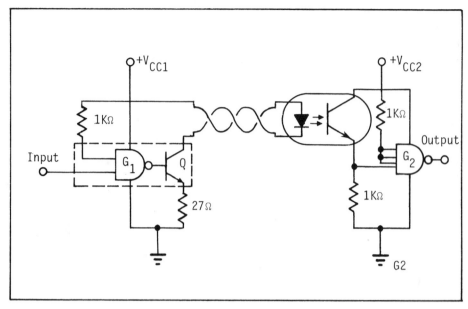

Figure 3.24 - Typical Interface System with ITT Circuitry, Twisted Pair and Optical Coupler.

coupled by a twisted pair line. Gate G_1 and transistor Q constitute the input stage driving the transmission line and emitter of the optical isolator. The LED requires about 20mA during *turn on*, which is well below the maximum current rating of the transistor. At the receiving end of the line, the phototransistor is coupled to a fast switching gate, G_2, for fast pulse generation.

Employing an optical isolator, as described above, provides isolation for both noise and high voltage. An isolation transformer or relay could accomplish the task, but it would not be as fast as the optical isolator. Also, a line driver and receiver combination could be used to eliminate the noise and increase the speed, but it would be very ineffective if there were high potential differences between the input and output. Interfacing microcomputers to comparatively high voltage input signals present in industrial process control applications can be accommodated via optical isolation. Digital input units containing optical isolators protect the microcomputer from standard as well as over-voltage conditions.

Although the current abundance of inexpensive microcomputers has made the computer control and monitoring of many industrial process functions economically feasible, most popular microcomputers do not interface readily with some of the control systems. For example, large industrial process monitoring relays and switches generate

widely varying high voltage pulses that must be conditioned before
interfacing with low voltage microcomputers. On/off states of large
24-Vdc or 110-Vac relays must somehow interface with the 5-V metal-
oxide-semiconductor or transistor-transistor logic level digital
transitions of the microcomputer. In addition, analog-to-digital and
digital-to-analog conversion interfaces must be provided for system
inputs and output.

As one solution of this problem, direct compatibility can be attained through use of digital input units (DIUS). Optical isolators
in the DIUS protect the microcomputers from damaging voltage transients, surges, and malfunctions that occur in harsh industrial environments.[6,16] In addition, optical isolation tolerates widely varying inputs and prevents the occurrence of ground loops.

3.4.5 Photo-Diode Optical Transformer

By connecting a pair of optical isolators in a positive-feedback
arrangement (Fig. 3.25),[17] an optically-isolated bilateral transmission device results that offers a number of performance benefits over
the conventional transformer. The transmission gain of the device corresponds to the step-up or step-down function of a transformer. However, this gain can yield bilateral amplification, if the current-
transfer ratio of the optical isolator is relatively large and the
collector capacitance of the coupler's internal output phototransistor
is very small.

Moreover, the input impedance of the new transmission device is
nearly independent of external load impedance. As a result, it can
provide impedance buffering, something the conventional transformer
cannot do. Also, when built with couplers having cutoff frequencies
of several kilohertz, the device can operate at frequencies of up to
a few hundred kilohertz. Applications are primarily in telecommunications, i.e., as a matching transformer and for suppressing impedance
variations in two-wire lines.

3.4.6 Interfacing Darlington Optical Isolators to TTL Logic

Through design and material improvements, Darlington optical
isolators are now as reliable and stable as visible emitters or any
other semiconductor device. Even the old problem of heavy interface
loading has been eased.

Packaging improvements in the new isolators allow the phototransistor and output transistor terminals to be brought out to separate
pins (Fig. 3.26). It is now possible to combine gain or bandwidth
networks to make an isolator compatible with logic family; it also

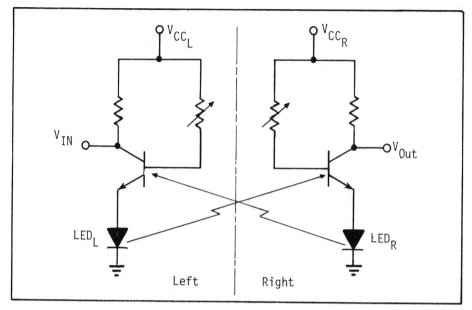

Figure 3.25 - Optical Transformer Provides Transmission Gain for Step-Up or Step-Down Function.

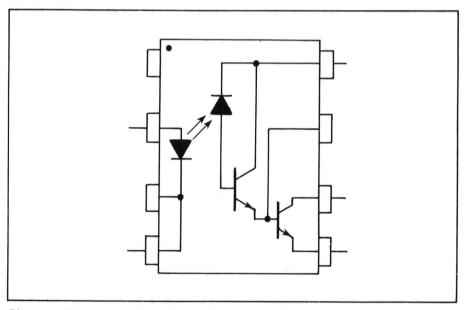

Figure 3.26 - Optical Isolator Package with Separated Input/Output Pins for Logic Interface.

takes fewer ICs to interface an isolator with other circuits. A few
simple ground rules are now given for hooking up isolators to various
types of logic.[4,18,19]

The primary design task when using a noninverting interface circuit (Fig. 3.27) is to determine the values of external resistors, R_1
and R_2 their values depending on the kind of logic used. First, however, the circuit is examined to see why this configuration is noninverting.

With a low level at the input, the internal LED is reverse-biased
keeping it off; therefore, the diode and Darlington that make up the
detector portion also remain off. This produces a high at the isolator output that is inverted by a NAND gate. Therefore, input and output are both low levels.

When the input pulse goes high, the LED and detector are turned
on, generating a low at the isolator output. The resulting high output of the NAND gate matches the input high.

In an inverting interface circuit, the external inverter is

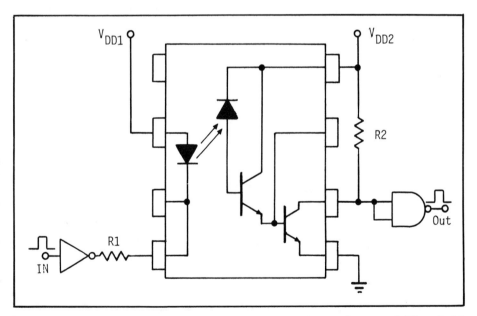

Figure 3.27 - Adding Resistors R_1 and R_2 to the Isolator of Fig. 3.26,
Completes a Noninverting Logic Interface. Resistor Values Depend on
the Logic Family in a Particular Application.

3.29

connected to the LED's anode instead of the cathode (Fig. 3.28). An input high turns off the LED and detector, giving a high at the isolator output, and a low at the NAND gate output. Of course, an input low turns on all devices in the isolator and results in a high output from the NAND gate.

Values for R_1 and R_2 can be calculated for several common types of logic using the equations and definitions of terms given in Table 3.1. Substituting the saturation and cut-off voltage levels for a given logic family into the equations, some important advantages of optical isolators become evident. It is found that an isolator requires a very low input drive current, and that its output current will be large enough to drive substantial loads.

3.4.7 Optical Isolators as Limit Switch Circuits

Optical isolators are very attractive in automation, where, as optical sensors, they can replace mechanical switches and operate quietly without requiring any force. They can be activated by a web

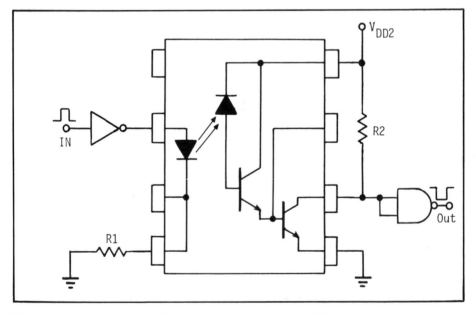

Figure 3.28 - An Inverting Logic Interface Requires the Same Resistors as in Fig. 3.27, but here R_1's Value Depends both on the Logic Family and the Circuit Configuration-Inverting or Noninverting, while R_2 is a Function of the Logic Only.

Table 3.1 - Equations for Calculating External Resistors
R_1 and R_2 of Optical Isolators Used as Logic Interface Circuits[18]

$$R_1 \text{ (NON-INVERT)} = \frac{V_{DD1} - V_{DF} - V_{OL1}}{I_F}$$

$$R_1 \text{ (INVERT)} = \frac{V_{DD1} - V_{OH1} - V_{DF}}{I_F}$$

$$R_2 = \frac{V_{DD2} - V_{OLX} \left(\alpha I_L + I_2 \right)}{I_L}$$

Where: V_{DDI} = INPUT SUPPLY VOLTAGE

V_{DD2} = OUTPUT SUPPLY VOLTAGE

V_{DF} = DIODE FORWARD VOLTAGE

V_{OL1} = LOGIC "0" VOLTAGE OF DRIVER

V_{OH1} = LOGIC "1" VOLTAGE OF DRIVER

I_F = DIODE FORWARD CURRENT

V_{OLX} = SATURATION VOLTAGE OF OPTICAL ISOLATOR

I_L = LOAD CURRENT THROUGH RESISTOR R_2

I_2 = INPUT CURRENT OF OUTPUT GATE

of paper or a foil vane attached to a meter movement.[3]

Figure 3.29 shows an optical limit switch of the reflecting type. The detector sees the reflected infrared light and provides a collector current proportional to the brightness of incident light. Since the detector is open to ambient light, it must be shielded from direct illumination or the detector will be saturated and not be able to respond to the light from the LED.

Another optical limit switch, a slotted type, is shown in Fig. 3.30. This sensor is designed to provide some degree of shielding and can function in normal room ambients.

The electronic circuits used with the optical sensors must be properly designed to match the performance of the photo-Darlington circuit in the detector.

3.31

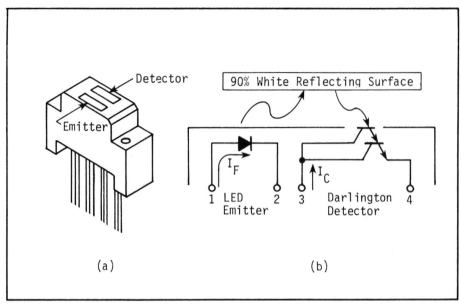

Figure 3.29 - Reflecting Object Sensor, a) Typical Package, b) Circuit Diagram with Reflecting Surface.

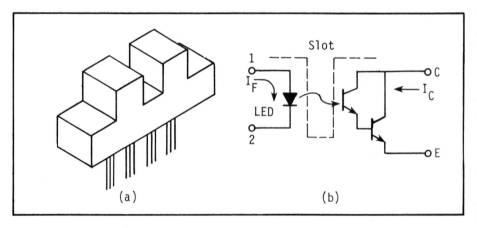

Figure 3.30 - Slotted Object Sensor, a) Typical Package Circuit and Slot Diagram.

A typical industrial application is shown in Fig. 3.31.[11] Here a circuit is required that can reliably and rapidly sense the presence of small parts on a wide moving belt and provide an output to 5-V logic. Cost of the control circuit, both initial and life cycle, is also important. Basically this is an interrupter module application. The relatively long distance that must be spanned and a high long-term reliability indicate that the LED should operate in a pulse mode. Speed of response then dictates use of a phototransistor detector biased from a very low source impedance.

Pulse operation permits use of the low-cost plastic interrupter module, thus eliminating the need for an expensive optic system and assembly alignment. Synchronous detection of the phototransistor output prevents leakage current, ambient light and temperature from degrading system performance.

A phototransistor coupler provides an ideal switch for synchronous signal detection. A programmable unijunction transistor (PUT) oscillator provides the least expensive pulse generator to drive both the interrupter and the coupler LEDs. A cascode-input amplifier, with heavy low frequency negative feedback, provides the required low source impedance bias for the interrupter phototransistor, as well as regulating the dc bias point to minimize ambient light, leakage and temperature effects. Three plastic-cased transistors provide the amplification. The amplifier output synchronously detected by PC_1, provides a negative output when the interrupter model's light path is unblocked. The system is fail-safe, since any malfunction will cause a zero or positive output signal. A simple two-transistor buffer amplifier with hysteresis provides the 5-V logic output, while permitting the use of a small detector capacitor to preserve system speed.

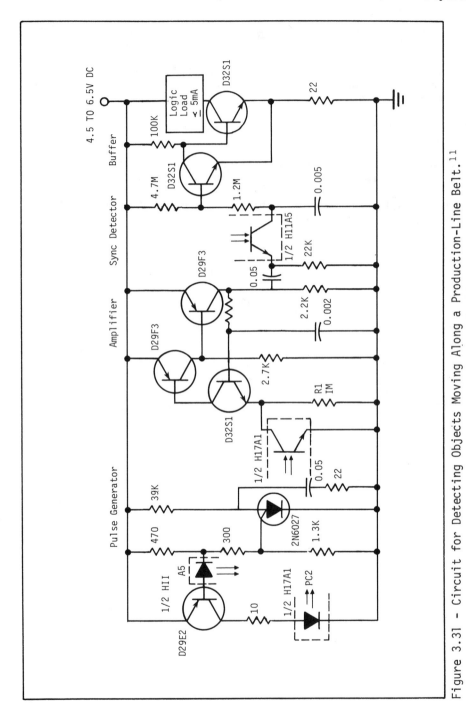

Figure 3.31 - Circuit for Detecting Objects Moving Along a Production-Line Belt.[11]

3.5 TESTING AND DEVICE SELECTION

To check the performance of an optical isolator against the manufacturer's specifications, a conventional transistor curve tracer can
be used. Most tests can be performed before the component is installed, thus saving many costly hours of trouble-shooting in the prototype,
production and test stages of manufacturing.

3.5.1 Testing With Curve Tracer

Consider a light-emitting diode and a photo-transistor in a
single package, as shown in Fig. 3.32 without the external circuitry
connected. The characteristics of the diode and the characteristics
of the transistor can be measured in the same way as for any other
diode or transistor. The two optical-coupling characteristics - ratios
of transistor *collector current* and *base current* to diode forward current - and the isolation can be checked by a Tektronix 577 curve
tracer. Other instruments can also be used in a similar manner. The
isolator may be connected to the curve tracer in the same way as a
standard diode or transistor. Since many optical isolators are
packaged in a six-pin mini-DIP flatpack, a dual-in line socket and
adapter allows easy connection of the device to the curve-tracer terminals. . As an alternative, a standard dual-in line IC socket, with
banana plugs wired to the terminal, can be used.

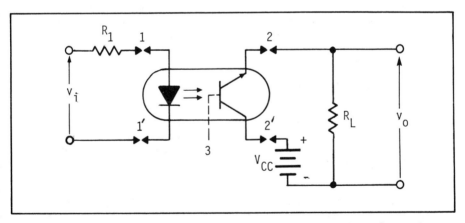

Figure 3.32 - Test Circuit for Measurement of Optical Isolator Performance.

Having access to leads 1-1', 2-2' and 3 (without the external circuitry connected) the following tests can be made:

- *Measurement of the dc collector current-transfer ratio* -- This is the ratio of dc collector current to the diode forward current. To check this value, pin 2' of the device must be connected to the collector terminal of the curve tracer, pins 1' and 2 to the emitter terminal, and pin 1 to the base terminal.

- *Measurement of base-current-transfer ratio* - This test can be done by removing the wire connected to the emitter of the isolator (pin 2) and connecting it to the base (pin 3). This change grounds the base and opens the emitter, thus allowing the collector base current to be measured.

- *Testing of isolation voltage* - To test the isolation voltage of the optical isolator, the curve tracer collector supply terminal can be connected to any one point on an element of the device, such as the diode, and one point on the transistor. By applying the specified voltage a check can be made for any leakage current.

- *Testing of the switching time* - For testing the switching time performance of the device, it is connected to the testing circuit as shown in Fig. 3.32, via terminals 1-1' and 2-2'. The LED can be operated in the pulsed mode and both the input pulse and output pulse can be observed on an oscilloscope to determine the actual switching time.

3.5.2 Evaluation Criteria

Even the best-made optical isolators can prove troublesome in high-volume applications if they are not specified and the quality of incoming devices is not controlled. Out-of-tolerance isolators may cause reduced or haphazard circuit performance, or outright circuit failure.

Five different types of optical isolator outputs are available – transistor, Darlington, SCR, diode/amplifier and diode/gate (Table 3.2). However, no matter which type is used, there are two basic ways to guarantee optimum performance;[11]

a) By identifying all the critical parameter-leakage, lifetime, bandwidth, coupling efficiency, speed, cost and specifying tolerance needed.

Table 3.2 – Comparison of Different Optical Isolator Types *

Type of Optical Isolator / Parameter	Transistor	Discrete Darlington	SCR	IC Diode/Amplifier	IC Diode/Gate
Efficiency	$CTR = I_C/I_F$ 10% to 100% min	$CTR = I_C/I_F$ 100% to 500% min	$I_{FT} = f\,(R_{GK})$ 20 mA max	$CTR = I_C/I_F$ 7% to 500% min	$CTR = I_O/I_F$ 600% min
Speed	t_{on}, t_{off} 1 to 5 μs	t_{on}, t_{off} 5 to 20μs typ	t_{on} 5 to 20 μs typ	t_{pd} 0.5 to 60 μs	t_{plh} t_{phl} 50 to 100 ns
Parasitics	C_{ISO} R_{ISO} 1 to 3pF, 10 to 100CΩ	C_{ISO} R_{ISO} 1 to 3pF, 10 to 100GΩ	C_{ISO} R_{ISO} coupled 100-500V/μs dv/dt 1 to 3pF, 10-1000GΩ	CMRV 20-60 Vrms @ 2MHz	$CMRV_H$ $CMRV_L$ 2-60Vrms @ 2MHz
Off Voltage	B_{VECO} B_{VCEO} typ. 5-7 min, 25V min	B_{VECO} B_{VCEO} typ 5-7V min, 25V min	V_{DM} V_{RM} 200-400 V min	$V_{KB(max)}$ $V_{CE(max)}$ 7 to 18 V	$V_{CC(max)}$ $V_{O(max)}$ 7 V
On Voltage	$V_{CE(SAT)}$ typ. 0.1 V	$V_{CE(SAT)}$ typ. 0.7 V	V_{TM} 1.3 to 1.5 V max	$V_{CE(SAT)}$ 0.1 to 0.2V typ.	V_{OL} 0.6 V max
Leakage	I_{CEO} 50-500 nA max	I_{CEO} typ. 100 nA max	I_{DM} typ. 50μA max	I_{CEO} 0.5 to 250μA max	I_{OH} 250 μA max
Maximum Power input or output	P_{do} typ. 400 mW max	P_{do} typ. 400 mW max	P_{do} typ. 400 mW max	P_{do} typ 100 mW max	N(FAN OUT) 8 max
Reverse Voltage (Min)	$V_{R(MAX)}$ 3 to 6 V	$V_{R(MAX)}$ 3 to 6 V	$V_{R(MAX)}$ 3 to 6 V	BV_R 5 V	BV_R 5 V
V_F (max forward volt)	1.5 to 2 V	1.5 to 2 V	1.5 to 2 V	1.7 to 1.8 V	1.7 to 1.8 V
Maximum power P_{dt} input or output	100 mW	100 mW	100 mW	35 mW	$I_{F(MAX)}$ 10 mA

* Reprinted with permission from Electronic Design, Vol. 23, No. 12, June 1975, copyright (Ref. 11) Hayden Publishing Co., Inc.

b) By controlling the critical factors through a combination of tests and vendor selection.

Since requirements vary with application, so will quality control methods. When special performance measurements are required, the following should be kept in mind:

- Most static parameters can be measured automatically and inexpensively.

- Dynamic parameters, high dielectric stresses and very low leakage currents require expensive manual testing.

- Destructive testing, which includes most burn-in mechanical, environmental and some high-stress dielectric testing, costs even more.

3.5.2.1   Evaluation of the Useful Coupler Lifetime

The light output of the LED in an optical isolator decreases as the LED is operated. This is a fact which is not normally expected of semiconductors. When an optical isolator is evaluated for light fall-off (loss of coupling efficiency), it would help to know that:

- Rates of fall-off vary widely from manufacturer to manufacturer.

- Liquid epitaxial processed material is much superior to any other process.

- Tight process control is required both in chip manufacture and in packaging to provide consistently good results.

- In a good device, stress-temperature and bias current must be increased greatly to produce appreciable acceleration of the phenomena.

- Short, high-current pulse operation causes much less fall-off of output than expected.

For a benchmark, a good product design will normally have less than 10% of the device population drop more than 20% in light output after 1000 hours of steady-state, maximum-rating testing. This extrapolates to about a 100-year half-life at maximum rating for 90% of the units.

3.5.2.2   Isolation

Isolation voltage and maximum dielectric capability are well specified when tested on a one-shot basis, but which of these should be used on a transient basis and which for steady-state? The answer depends on the voltages at which corona initiates and extinguishes. Since the corona is not easily detected and bears no strong correlation with the maximum spec value in different mechanical designs, the designer must lab test it himself.

Environmental effects can normally be handled with straight forward design procedures such as good temperature control, clean circuit-board construction and careful circuit layout. There are also two commonly unforeseen problems. First, leakage current increases faster than expected for semiconductor devices since a base-to-emitter resistor usually is not used in transistor and Darlington coupler applications. Second, coupling temperature coefficients are not always the same for different manufacturer's devices at low current levels. Both of these potential problems are easily avoided when known.

In logic applications, when a coupler is TTL-compatible, it must operate with standard tolerance pull-up resistors, over standard supply-voltage tolerances, over required temperature ranges and at worst-case logic levels.

3.5.3 Comparison of Various Signal Coupling Devices

Optoelectronic isolators and couplers are replacing the slow and bulky electromechanical devices that previously isolated delicate circuitry, especially in telecommunications, which constitutes the devices' biggest market. As these devices mature in complexity and capability, they could become commonplace in all kinds of digital equipment. With the increase in distributed processing, optical isolators have taken on a new importance to the computer industry.

In summary, optical isolators are extremely versatile devices and can provide better solutions to many system problems than other competing components shown in Table 3.3.[2]

Table 3.3 - Summary of
Properties of Signal Coupling Devices

Device	Advantages	Disadvantages
Optical Isolator	Economical Solid State reliability Medium to high speed signal transmission DC & low frequency transmission High voltage isolation High isolation impedance Small size DIP Package No contact Bounce Low power operation	Finite ON Resistance Finite OFF Resistance Limited ON state current Limited OFF state voltage Low transmission efficiency (Low CTR)
Relays	High power capability Low ON resistance DC transmission High voltage isolation	High cost High power consumption Unreliable Very slow operation Physically large
Pulse Transformers	High speed signal transmission Moderate size Good transmission efficiency	No DC or low frequency transmission Expensive for high isolation impedance or voltage
Differential Line Drivers and Receivers	Solid state reliability Small size DIP package High speed transmission dc transmission Low cost	Very low breakdown voltage Low isolation impedance

3.6 REFERENCES

1. Filichowski, W.; *LED Photoelectrics;* Digital Design, March 1977.

2. Smith, G.; *Applications of Opto-Isolators;* Appnote 2, Litronix Inc., Ca. 1971.

3. Monsanto; *Opto-Isolator Application Notes;* Monsanto Company, Palo Alto, Ca. 1973.

4. Optoelectronics Application Manual, Hewlett Packard Optoelectronics Division, 1977.

5. Tenny, R.; *Positive Feedback Speeds up Low-Cost Opto-Isolator Response;* Electronic Design 9, April 26, 1978.

6. Teeple, C.R.; *Digital Input Units Isolate Microcomputers from Industrial Level Voltages;* Computer Design, November 1978.

7. Teschler, L.; *Modular I/O for Microcomputers;* Machine Design, November 9, 1978.

8. Texas Instruments Application Report; Optically Coupled Isolators in Circuits; Texas 1970.

9. Heftman, G.; *Optoelectronic Components Sparked by Faster Switching, Higher Isolation;* Electronic Design 19, September 13, 1979.

10. Olson, H.; *Control a Latching Relay from TTL Inputs with a Dual Photo-SCR Coupler;* Electronic Design 2, January 18, 1979.

11. Sahm, W.; *Get to Know the Optocoupler;* Electronic Design 12, June 7, 1975.

12. Waaben, S.; *High Performance Optocoupler Circuits;* Int. Solid-State Circuits Conf. Dig. Tech. Papers, Vol. XVIII, 1975.

13. Vettiger, P.; *Linear Signal Transmission with Optocouplers;* IEEE Journal of Solid-State Circuits, Vol. SC-12, No. 3, June 1977.

14. El-Diwany, M.H., et al; *Piecewise Linear CAD Model for Avalanche Photodetectors;* Proc. IEEE, Vol. 67, No. 8, August 1979.

15. Hayes, S.; *Simple Analog Divider Uses Optical Coupling in its Feedback Loop;* Electronic Design 13, June 21, 1978.

16. RCA; Optically-Coupled Isolator; Lancaster, Pa. 1975.

17. Mattera, L.; *From Japan: Optical Transformer, Photo-Process for Large-Scale Hybrids;* Electronic Design 13, June 21, 1979.

18. Otsuka, W.; *Increased Reliability and Higher Gain Spurs New Optoisolator Applications*; Electronic Design 26, December 20, 1978.

19. Texas Instruments Electronic Series; *Optoelectronics: Theory and Practice*; McGraw-Hill Book Co., 1978.

20. Grossman, R.; *Versatile Optoisolators, Couplers Move Into New Applications*; EDN, April 5, 1979.

CHAPTER 4

FIBER OPTIC COUPLING AND CONNECTIONS

All optical systems use the same *type* of components as the ones encountered in microwave systems. In a practical sense, they perform the same, but instead of electrons flowing through a coaxial cable, photons (radiant energy) propagate through some medium - either air, or an optical fiber. Transmitters, receivers, and repeaters are still used. In fact, internally, these devices are nearly identical to microwave components. The reason lies in the young and growing field of fiber optics.

4.1 INTRODUCTION

Fiber optics are moving from research into production and into the marketplace. Today several technologies - optical, electrical, semiconductor, glass, and plastic (in the form of unique glass and quartz fiber optics, lasers, light emitting diodes, and advanced plastics) - converge to create a practical and economical new optical waveguide system for data transmission.[1,2] Fiber data links are now being built into off-the-shelf equipment.

In fiber optic links, specially designed connectors and/or optical couplers are used. The factors introduced by the connector system include lateral core misalignment, angular misalignment, end separation, end preparation quality, cleanliness and reflections.[3] Splices for joining lengths of fiber cables in the field must provide low-loss, quick, permanent connections which are small and rugged. A special class of couplers is needed to perform signal distribution in multiterminal (bus) data communications systems.[4]

When choosing components for a fiber optic system, be sure that the selections are wavelength compatible. For optimum system performance, the source's emission-peak wavelength should match the valleys in the fiber's attenuation-wavelength characteristics. Additionally, the detector must be responsive in this wavelength range.[5] Figure 4.1 depicts some representative components, including diode sockets, connectors and connecting cables.

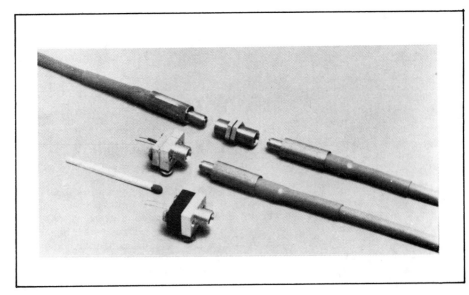

Figure 4.1 - Representative Optical Components: Cables, Connectors and Diode Sockets (*Courtesy of SIEMENS*).

4.2 DEFINITIONS AND DESCRIPTIONS

The special features of fiber optics and their associated components make it necessary to describe some of the properties of optical cables in terms different than those used for conventional cables with metallic conductors. These definitions and descriptions,[6,7] will help the reader to better understand the material of this chapter.

- *Fiber Optics*

    Fiber optics (FO) is the technology of passive guidance of optical radiation (rays and waveguide modes) along optical conductors.

- *Step Index Fiber*

    A fiber type consisting of two circularly symmetrical coaxial elements of homogeneous but differing refractive indices where the outer element (Cladding) is of lower refractive index than the inner element (core).

- *Graded Index Fiber*

    A fiber in which there is a variation of refractive index across the fiber core. The purpose of this variation is to equalize group velocities of the various propagating modes, thereby improving communication bandwidth.

- *Single Mode Fiber*

    A fiber waveguide through which only one mode will propagate.

- *Numerical Aperture (NA)*

    The numerical aperture describes the behavior of an optical cable when coupled to light sources, detectors or other optical components (e.g., in splices or connectors). the numerical aperture is defined as the sine of the angle between the fiber axis and an element of the surface of the cone enveloping the entire light emerging from the fiber core after a fiber length of two meters.

- *Terminations*

    The small diameter of fiber optic cables calls for precision-made connectors. In some cases, the hole in the connector through which the fiber fits is less than 1.5 mm

4.3

in diameter. Once the connector is installed on the cable, the surface must be ground and polished to provide a very precise, mirror-like surface. Any imperfection in the surface polish causes increased coupling losses. Very accurate alignment of the connectors is also required.

- *Bandwidth and Loss*

     Fiber optic cable presently available for applications can have bandwidths up to 3.3 GHz. Cables used in data and communications links typically have BW products of 200 to 400 MHz·km and exhibit signal losses typically less than 10 dB; in some cases, just 1 to 3 dB/km. Because the bandwidth of fiber is so enormous, it is unlikely that an installed cable would have to be replaced because of increased traffic. Also, because of the low loss of some fiber cables, fewer repeater or amplifier stations would be required for long distance runs. Wire systems, for example, typically require repeaters approximately every 400 meters; fiber systems are now operating over distances of up to 8 kilometers without repeaters.

- *EMI Immunity*

     Because fiber optic cables are made of glass and plastic, they are nonconductive, and do not pick up electromagnetic noise. This is a big advantage because it permits data to be transmitted without noise-induced error. Using fiber optics, it is possible to totally eliminate the common problem of crosstalk or cross-coupling.

- *Isolation and Security*

     Fiber optic cables are totally dielectric providing complete isolation between the interfacing equipment. Electrical isolation approaching 100 dB can be achieved.

- *Size and Weight*

     Because of its small size, fiber optic cable can be installed in areas where conventional wire cables will not fit. The cable's relative lightness reduces installation costs, and in areas such as airborne applications, can make a significant change in range and payload capabilities.

4.3 TRANSMISSION PROPERTIES OF FIBER OPTICS

There are three basic fiber optic types: (1) multimode-step index, (2) multimode-graded index and, (3) single mode-step index, as shown in Fig. 4.2.[4] They are distinguished primarily by the profile of the index of refraction across the fiber's cross-section. The *index of refraction (n)* is one of the main parameters of an optical fiber. Classically, for both glass and plastic, the index of refraction is defined as n = c/v, where c is the speed of light in vacuum and v its speed within the material. An optical fiber's *index profile* refers to the way its index varies as a function of radial distance from the fiber's center.

4.3.1 Multimode - Step Index Fiber [8]

The simplest way to consider transmission over an optical fiber is to think in terms of total reflection in a medium of refractive index n_1 at the boundary with a medium n_2, where n_1 is greater than n_2. This is the case of a typical multimode fiber such as shown in Fig. 4.3, which would have a circular core of diameter d and uniform refractive index n_1 surrounded by a cladding layer of refractive index n_2. Light launched into the core at angles up to θ_1 will be propagated within the core at angles up to θ_2 to the axis. Light launched at angles greater than θ_1 (as shown by the broken line in Fig. 4.3) will not be internally reflected, but will be refracted into the cladding or possibly even out of the cladding into the air at the second boundary if the launching angle is large enough, and n_1 and n_2 are small enough. The maximum launch and propagation angles are given by the numerical aperture NA:

$$NA = \sin\theta_1 = n_1 \sin\theta_2 = \left(n_1^2 - n_2^2\right)^{1/2} \qquad (4.1)$$

Most fibers are constructed with a core/cladding index discontinuity of a few percent or less, i.e.,

$$n_1 = n_2(1 + \Delta) \qquad (4.2)$$

where,

$\Delta \ll 1$. Substituting Eq. (4.2) into Eq. (4.1) yields:

$$NA = \sin\theta_1 = \left(n_1^2 - n_2^2\right)^{1/2} \approx n_1\sqrt{2\Delta} \qquad (4.3)$$

Since this is the equivalent of an electromagnetic waveguide propagation, only certain modes, which may be regarded as rays corresponding to specific quantized values of θ_2, can propagate. The number of modes N for light of wavelength λ is given by:

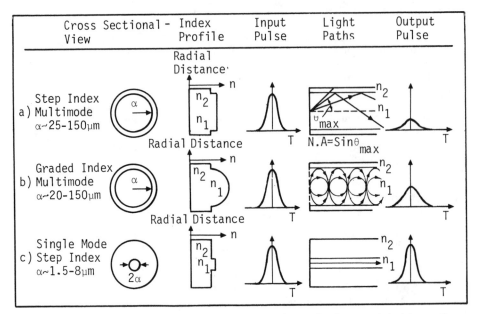

Figure 4.2 - Types of Optical Fibers - Schematic Representation of Cross Sections, Index of Refraction, Distributions and Optical Ray Paths. [4]

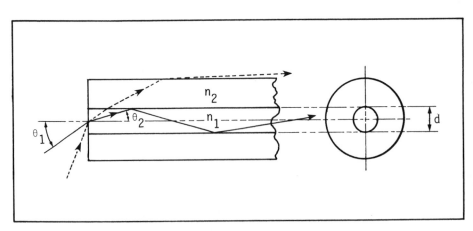

Figure 4.3 - Ray Paths in a Multimode-Step Index Fiber Optic.

$$N \approx \frac{1}{2} \left(\frac{\pi d NA}{\lambda} \right)^2,$$ (4.4)

where, d is the diameter of the core.

4.6

Thus, for a given combination of refractive indices, as the diameter of the core is reduced, fewer modes will propagate. When eventually the diameter becomes of the same order of magnitude as the wavelength of light, only a single mode will propagate.

Considering the maximum difference in the path length for a ray propagating parallel to the axis and a ray propagating at the maximum angle θ_2, it is possible to calculate the pulse spreading due to the different times taken by the energy propagation along the two different paths. Thus, the bandwidth capacity of a high NA guide is less than that of one able to support fewer modes.

4.3.2 Multimode - Graded Index Fiber [4,8,9]

In the construction of a multimode graded refractive index profile (Fig. 4.2b) the energy in the outer modes spends the longer time in the lower refractive index medium. Thus, the rays with the longer distance to go, spend more of their time in the medium with the higher propagating velocity. This tends to give the same delay for different modes of propagation and thus enables the lower dispersion to be obtained with higher numerical aperture than would be possible in the multimode fiber with its sharp change in refractive index.

From geometric optics it is found that light incident on the fiber core at position r will propagate as a guided mode only if it is within the local numerical aperture NA(r), at that point. The local NA(r) is defined as:

$$NA(r) = NA(0)\sqrt{1-(r/a)^{\alpha}}, \; r \leq a \qquad (4.5)$$

where $NA(0) = n_1\sqrt{2\Delta}$, a is the core radius, and α is the graded index power law coefficient with a value close to 2, depending on the exact composition of the fiber core and the operating wavelength.

For the graded refractive index fiber with α close to 2, the number of modes that will propagate with respect to Eq. (4.4), will be:

$$N = \left(\frac{1}{4}\right)\frac{\pi d(NA)^2}{\lambda} \qquad (4.6)$$

For a very low number of modes the approximation breaks down and the condition for a single-mode fiber is:

$$N = \frac{\pi d_s(NA)}{\lambda} \leq 2.4 \qquad (4.7)$$

where d_s is the core diameter for single-mode operation at wavelength λ.

4.3.3 Monomode - Step Index Fiber

As can be seen from Fig. 4.2c, in the single mode fiber a very
large part of the energy is carried within the cladding dielectric in
the evanescent field surrounding the core. As the core diameter is
further reduced below the maximum permitted value given by Eq. (4.7),
the single mode will still propagate, but it is less tightly coupled
to the core and more of the energy is transmitted in the cladding.
By definition, no dispersion is introduced in a single-mode optical
guide.

4.3.4 Dispersion and Attenuation Mechanisms [4,5]

Dispersion is the undesirable effect of the broadening of optical pulses caused by the lengthening of rise and fall times as the
pulse travels along the fiber. Signal distortion is due to two types
of fiber dispersion. Their relative importance depends upon wavelength and the fiber type. One type is the *multimode (or intermode)
dispersion* that occurs due to the variation of group velocity with
mode. Assuming that Δ is the percentage difference of the core-clad-
ding refractive indices, the detectable pulse rate will be:

$$R_s = 18/\Delta \text{ Mbit.Km/s} \tag{4.8}$$

where subscript s refers to step-index. Differential mode attenuation can enhance this by a factor of 2 or more, while mode mixing,
which usually increases scattering attenuation, leads to a sublinear
dependence of pulsewidth with fiber length L. This may be expressed
in terms of an equilibrium length, L_s, below which the pulse rate
varies as L^{-1} and above which it varies as $(L_sL)^{-1/2}$. Alternatively,
the index on L changes continuously from -1 to -1/2 for short to long
lengths, respectively.

More reduction in multimode dispersion occurs with graded-index
fiber optics, where the estimated[10] maximum pulse rates are:

$$R_g = 20/\Delta^2 \text{ Gbit.Km/s} \tag{4.9}$$

where subscript g refers to graded-index. As with step-index fiber
optics, mode-mixing and differential attenuation effects can reduce
these values.

The second type is the *chromatic dispersion* that decreases as
source monochromaticity increases and is due to both material and
modal effects.

Another fiber characteristic of interest is *attenuation*, particularly in low-loss fibers. Attenuation is due in part to intrinsic absorption by the atoms which constitute the fiber, absorption by atomic defect color centers, and by extrinsic impurity absorption. The other contributing factor to attenuation is *scattering* due to index and fiber shape inhomogeneities.

4.3.5 Spectral Matching of Fiber Optics to Sources and Detectors

When choosing components such as emitting diodes, optical fibers and detector diodes, it is especially important to ensure that they are matched to each other spectrally. The spectral maximum.of the emitting diodes should agree with an attenuation minimum of the fiber and the maximum of spectral sensitivity of the receiving diode[11] (Fig. 4.4). This is particularly important when the user does not obtain complete fiber optic communications systems, but prefers to build up his own system from components of various manufacturers, requiring the components to be carefully matched.

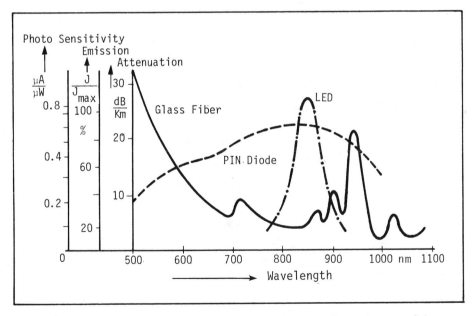

Figure 4.4 - Spectral Compatibility of Emitters, Detectors and Low-Loss Fibers (*Ref. 11*).

The trend now in fiber optic technology is the development of optical fibers for operation in longer wavelengths. The goal is to attain lower attenuation levels. Fibers having attenuation levels of under 1 dB/Km at wavelengths of 1.3 micrometers are now moving from the laboratory into production. The development of ultralow-loss optical fibers operating in the wavelength range of 1.2 to 1.6 micrometers, together with the realization of suitable, high-power, high-speed light-emitting diode sources, will permit the design of long-haul, high-data-rate links using LEDs instead of lasers.

4.4 SOME BASICS OF OPTICAL FIBERS AND CABLE PREPARATION

Optical fibers can be made from either glass or plastic. Glass fibers have the lowest losses (as low as 5 dB or less per Km) and widest bandwidths (as much as 400 MHz), but they are the most expensive. Plastic fibers have losses on the order of 100 to 1000 dB, and bandwidths of tens of megahertz.

The unprotected glass fiber is wholly unsuitable for practical use. It is necessary, as with copper conductors and waveguides, to form the fibers into a cable for use in other than laboratory environments. Optical communication cables use either fiber bundles or single fibers to carry information; the trend, however, is toward single fibers. Fiber bundles generally are used to distribute random light for display or illumination.

4.4.1 Optical Fiber Drawing Techniques [12]

A number of methods have been used to fabricate moderate-to-low loss optical fibers. Table 4.1 lists certain processes and resultant fiber characteristics. Very low-loss fibers need impurity levels of some transition metals of 10-50 ppb. Also, other quality restrictions to limit scattering are equally stringent. At this point, Corning's doped-deposited silica (DDS) technology should be mentioned, which first attained the glass purity required for low optical loss. There are two versions of the DDS process, which is a vapor oxidation process. The first is outside vapor-phase oxidation (OVPO) and the second, inside vapor-phase oxidation (IVPO).[12] A representative arrangement for the OVPO process is shown in Fig. 4.5.

Table 4.1 - Optical Fiber Fabrication Processes

PROCESS	GLASS TYPE	FIBER LOSS (dB/Km)
Rod in Tube	Crucible melt	500-1000
Phasil	Leached Borosilicate	10-50
Compound melting	High SiO_2 (typically Pd-doped)	5-20
Plastic Clad	SiO_2 core Plastic cladding	5-50
DDS		
IVPO	High SiO_2	≤ 10
OVPO	High SiO_2	≤ 10

Fibers are drawn by placing a blank of glass into a furnace, and subsequently heating the tip of the blank until the glass softens. During drawing, the core and cladding glass maintain their respective

4.11

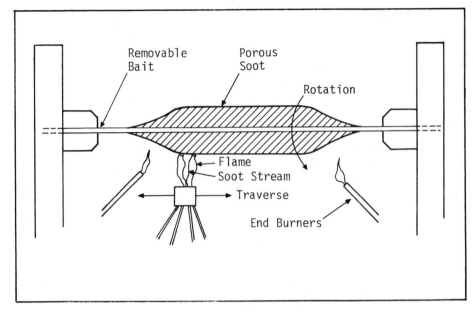

Figure 4.5 - Outside Vapor-Phase Oxidation Process.

geometric relationships, even though diameter attenuations are as
great as 300:1. The core-to-cladding ratio and the refractive index
profile of the core in blank form are faithfully reproduced in the
fiber. The critical fiber properties are: (1) optical, (2) dimension-
al, and (3) strength.

Various other methods of fiber optic construction are described
extensively in References 13 and 14.

4.4.2 Cable Designs [15]

There is some similarity between the various parts of a conventional copper cable and a fiber optic cable as shown in Fig. 4.6.

First, some standard optical cable designs are considered which
have been proven as indoor and outdoor cables for tray, duct and
underground installation.[15] Figure 4.7a shows the cross-sectional
diagram of an indoor-fiber cable. A layer of reinforcing fiber glass
yarn is spun around the buffer jacket and protected by a polyurethane
(PUR) jacket. This light-duty cable is designed for use in a variety
of environments over moderate installation distances. The fiberglass
strength member is used to impact a degree of longitudinal stiffness
to the cable. The tightly-extended PUR sheath is still flexible and

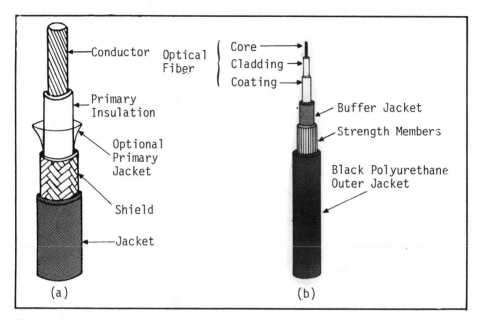

Figure 4.6 - Parts of Typical Conventional Copper Cable (a) and Fiber Optic Cable (b).

provides, together with the glued fiber-glass, a longitudinal stiffness for a broad operating temperature range (-30° to 50°C). This cable can be reinforced by an additional layer of Aramid-yarn (Fig. 4.7b) which is covered by an outer PVC sheath. For a two-fiber cable construction, two subunits as shown in Fig. 4.7c are used and reinforced by a layer of Aramid-yarn. The material of the outer sheath of this flat cable is either PVC or PUR.

In order to meet special requirements, the standard cable design can be modified easily. Some of these modifications are briefly mentioned below:

- *All dielectric cables*. These are desirable in environments with a high probability of lightning. The control element in these cables can be replaced by a fiberglass reinforced plastic element.

- *Flame retardant cables*. These are required for building installations and for nuclear power plant applications. They must meet the IEEE-383 flame test.

- *Copper wires incorporated into fiber cables*. These are needed where repeaters have to be powered.

4.13

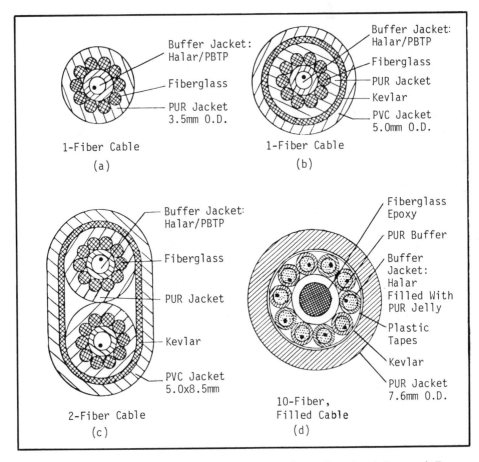

Figure 4.7 - Structures of Fiber Optic Cables: Standard One-and Two-Fiber Cables (a, b and c) and Modified all Dielectric 10-Fiber Cable Containing Buffer Jackets Filled with PUR-Jelly (d).[15]

- *Filled cables.* Such fiber optic cables are recommended in all cases where the presence of water due to damage to the cable jacket, together with freezing temperature, is to be expected.

- *Rodent protection for underground cables.* Rodent protection may be achieved by an additional outer PUR sheath or a steel tape armoring.

- *Direct burial cables.* These are exposed to high pulling forces and additional mechanical loads.

- *Aerial cables.* They are made with design options including concentric self-supporting, figure-8 and lashed to a messenger cable.

- *Pressure tight cables.* These are suitable for underground nuclear testing

Table 4.2 on the following page, shows some representative types of fiber optic cables and their characteristics.[16]

4.5 CONNECTORS AND COUPLERS

Due to rapid advancement of optical fiber technology, the availability of connectors and couplers for such fibers was lagging until recently. Now, however, numerous connector and coupler types have been announced and some of them are already available as off-the-shelf items.

The object in connecting two optical waveguides is to maximize the interconnection of their mode volume. This interconnection may be made less than perfect through factors which are intrinsic to the fiber being connected and others which depend on the quality of the connector system.

4.5.1 Connectors and Joining Techniques

The factors introduced by the connector system include lateral core misalignment, angular misalignment, end separation, end preparation quality, cleanliness and reflections. Thus, the problem of connector system design reduces to the achievement of cost-effective control and minimization of these fiber-extrinsic factors. Special areas to be observed by the connector designers and fiber optic system designers include:[17]

- The overall concept should be as simple as possible and still achieve the target of low connector loss. Parts count should be minimized.

- The requirement on manufacturing precision should be low. Parts should be producible by methods which are conventional or nearly so. Tolerances on diameters and concentricities should be within normal practices.

- The size of the connector should be within a range in which bulk is minimized and yet tedious assembly and handling methods are not needed.

4.15

TABLE 4.2   REPRESENTATIVE SAMPLINGS OF OPTICAL FIBER CABLES[16]

MFG	DESIGNATION	NUMBER OF FIBERS	CABLE TYPE	FIBER TYPE	ATTENUATION (dB/km)	BANDWIDTH (-3 dB) OR DISPERSION (3 dB WIDTH)	FIBER CLADDING/CORE DIAMETER (μm)	CABLE DIA (mm)	CABLE WEIGHT (kg/km)	MIN BEND RADIUS (cm)	TENSILE STRENGTH (kgf)	COMMENTS
DUPONT	PFX-S120R	1	T	PS	50		600/200	2.4	6	0.3	65	
ITT	ESM-6-GG(7)	7	T	GG	6	2.5 nSEC/km	125/50	6.4	30	5	100	Plastic-Clad
	ESM-6-GS(7)	7	T	GS	6	15 nSEC/km	125/50	6.4	30	5	100	Fiber Cables Available
	S-2-GG(2)	2	T	GG	6	2.5 nSEC/km	125/50	2.5 x 5	-	-	-	
SIECOR	Premium	6	L	GG	6	400 MHz	125/62	7.3	45.5	5	40	1-,8 and 10 - Fiber Cables Available
	Standard	4	L	GG	10	200 MHz	125/62	6.5	33	5	40	
	Premium	2	L	GG	6	400 MHz	125/62	5 x 6.4	26.6	5	40	
TIMES	GPI/SA7-90	1	L	GS	7	50 MHz	125/90	2.8	9.5	2.8	89	2-,4-,6 and 10 - Fibers Also Available
	GP3/GA10-90	3	L	GG	10	300 MHz	125/90	5.6	22	5.6	89	
VALTEC	LD-SG04-G1	1	-	GS	<10	<1 nSEC/km	100/5	3	-	-	-	Single Mode
	MD-PS10-02	2	L	PS	<15	10 MHz	430/250	5 x 9	-	2	100	
	XD-MG05-06	6	L	GG	<5	400 MHz	125/62	15	-	-	900	Telephone Cable

CABLE TYPE: T-Tightly Bound: L-Loosely Bound

FIBER TYPE: PS-Plastic Clad, Step Index: GS-Glass Clad, Step Index: GG-Glass Clad, Graded Index

- Low connector loss should be achieved in all connections between randomly selected parts and should be preserved on repeated disconnections and reconnections of the same set of parts.

- The connector should offer inherent protection of the fiber and inherent connector strength in terms of strain relief, shear resistance, etc.

- Design effort should be directed not at just a connector or the terminations, but at a complete connection system.

- The connection scheme should offer low component costs, in small volumes as well as in large.

The key design problem in any single fiber connector is to align the central axis of one fiber with the central axis of the other. Since each fiber has its own characteristics of size, numerical aperture, etc., measurements of the effect of end separation, axial alignment and other factors must be considered.

Figure 4.8 shows some representative techniques that are used for connecting fiber optic cables[18] and Table 4.3 is a list of selected single-fiber optical connectors.[16]

Methods of coupling light sources or light detectors into optical cables fall into four categories, as shown in Fig. 4.9.[3,19,20]

Another problem that engineers and maintenance people are faced with in field installation conditions is splicing of fiber optics.[22] Bearing in mind that the typical diameter of an optical fiber core is less than 100 μm, and considering the problems of precision alignment in field situations, the requirements for good fiber optic splicing pose a major challenge.

Here, two main techniques are explored: *mechanical splicing and fusion splicing*. Using *mechanical splicing* equipment (Fig. 4.10a) permits easy, permanent splicing of optical fibers. The mechanical splicer can be provided with other parts such as a *buffer cutter, splice connector, and crimping pliers* and does not require an energy source for splicing. Mechanical splicing, however, usually requires index-matching epoxy to be mixed on site, and sufficient time allowed for curing after application.

Considerable work has been done on fusion splicing[21,22] The fusion or *hot splice* method uses an electric arc or other heat source to fuse two fiber ends together after they have been accurately aligned. A *fusion splicing set* is shown in Fig. 4.10b. The power and duration of the arc are adjustable in this set for specific sizes and types of fiber, and the settings are maintained automatically.

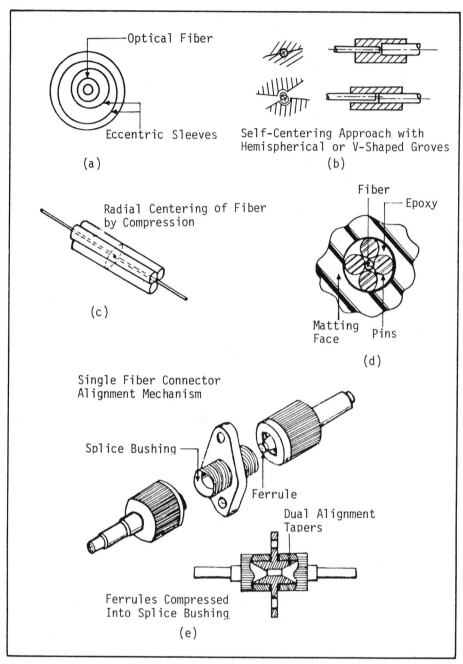

Figure 4.8 - Various Techniques for Optical Cable-to-Cable Connections.

Table 4.3 - Examples of Single-Fiber Optical Connectors[16]

MFG	DESIGNATION	NUMBER OF FIBERS	FIBER DIA (μm)	INSERTION LOSS (dB)	COMMENTS
AMP	OSC 22658	2	400	2	FOR DUPONT PFX-P SERIES OF PLASTIC-FIBER CABLES
AMPHENOL	906 SERIES	1	125	1.5-2	TERMINATES SEVERAL MFGS SINGLE-FIBER CABLE
	905 SERIES	1	600	3	PRIMARILY FOR BUNDLES BUT ACCOMMODATES DUPONT PFX S120R PLASTIC-CLAD SILICA FIBER
CANNON	UNILUX/FOS	1	100~325	2	ACCOMMODATES STRENGTHENED OUT-DOOR-ENVIRONMENT CABLES
ITT	MULTIWAY	4 OR 8	50~200	2	FACTORY ASSEMBLY OF CONNECTORS TO CABLE RECOMMENDED
T&B/ANSLEY	998-100	1	125	1.5	CABLE STRAIN RELIEFS AVAILABLE. FIBER CLEAVING AND ALIGNMENT TOOLS AVAILABLE
	SPLICE	1	125	-	NARROW-PROFILE SPLICE FOR PERMANENT CONNECTORS USABLE WITH INDEX-MATCHING EPOXY
	998-500	2	400	3-4	FOR DUPONT PFX-P SERIES OF PLASTIC-FIBER CABLES

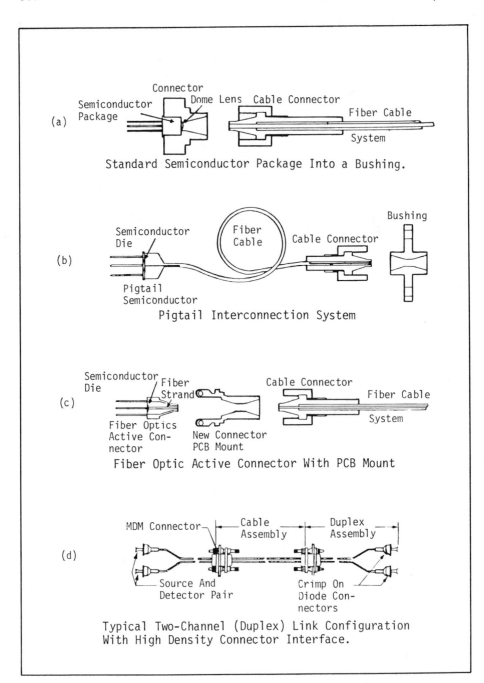

(a) Standard Semiconductor Package Into a Bushing.

(b) Pigtail Interconnection System

(c) Fiber Optic Active Connector With PCB Mount

(d) Typical Two-Channel (Duplex) Link Configuration With High Density Connector Interface.

Figure 4.9 - Methods of Coupling Optical Cables to Light Sources or Light Detectors.

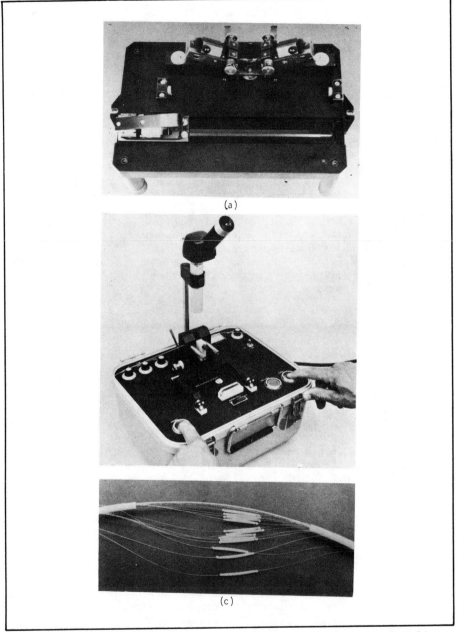

(a)

(c)

Figure 4.10 - Fiber Optic Splicing Equipment. a) Mechanical Splicer
(*Courtesy of Siemens*). b) Fusion Splicing Set (*Courtesy of NORTHERN
TELECOM.*). c) 12-Fiber NORTHERN TELECOM Optical Fiber Cable with all
12 Fibers Spliced.

The set is provided with a diamond cleaving tool for edge cutting of
the tiny fibers with great angular precision. Figure 4.10c shows a
12-fiber NORTHERN TELECOM optical fiber cable with all 12 fibers
spliced.

4.5.2 Minimization of Fiber Optic-Connector Losses [23,24,25]

In considering coupling of fiber optics, the losses that must be
taken into account can be broken down into two general classifications,
intrinsic and *connector-related*. *Intrinsic* losses are inherent in
the fibers; there is not much the designer of the connector can do
about them except include them in the tabulation of overall losses
through the coupling.

Coupling losses inherent in a pair of fiber optics primarily
derive from the fibers' differences with respect to three parameters
(1) radius, (2) numerical aperture, and (3) index of refraction.
Also, differences in core diameter of the connecting elements - source
port to receiving fiber, or transmitting fiber to detector - can cause
losses. These losses are given by:

$$dB = 20 \log\left(\frac{D_r}{D_t}\right) \quad D_r > D_t \qquad (4.10)$$

$$dB = 20 \log\left(\frac{NA_r}{NA_t}\right) \quad NA_r > NA_t \qquad (4.11)$$

where the subscripts r and t refer to receiving and transmitting ends,
respectively.

Another intrinsic loss for bundle fibers involves the *packing
fraction (p.f.)*. Since each individual fiber is circular in cross-
section, a dead area exists between fibers even when consolidated into an efficient hex pack. The ratio of the active core area at a
bundle end face (termination) to the total termination is the packing
fraction. Thus, for an m-fiber bundle:

$$p.f(dB) = 10 \log \left[(\Sigma_1^m \, d^2)/D^2 \right] \qquad (4.12)$$

where d is the fiber core diameter and D the termination diameter.

The *Fresnel* loss or reflection loss is also considered intrinsic.
It occurs due to the differences in refractive indices at a fiber-to-
fiber interconnection. For light propagating through the glass-air-
glass interface, the loss will be:

$$dB = 20 \log (1-\left[(n_g-n_a)/(n_g + n_a)\right]) \qquad (4.13)$$

where n_a is the air index and n_g the glass index. It should be noted that a fluid with an index of refraction matching that of the fiber core between fiber ends will eliminate the Fresnel loss. A problem exists, however, in cleanliness and retention of the fluid on repeated matings; therefore, use of fluid does not provide a reliable separable connector.

End finish of the fibers and fiber distortion (to a certain extent) can be minimized by using careful cutting techniques.

The remaining parameters that the connector designer has to work and optimize are lateral or axial misalignment, end separation, and angular offset. These cause extrinsic losses. In a lateral misalignment (Fig. 4.11a) the loss is:

$$dB = 10 \log\left[1-2L/(\pi D)(1-\delta^2/D^2)^{\frac{1}{2}} - (2/\pi)\sin^{-1}(\delta/D)\right] \qquad (4.14)$$

End separation loss, a function of numerical aperture and gap distance (Fig. 4.11b), is given by:

$$dB = 10 \log \left[(D/2)/(D/2 + g \tan\theta_c)\right] \qquad (4.15)$$

where,

$$\theta_c = \sin^{-1}(NA/n) \qquad (4.16)$$

An angular loss (Fig. 4.11c) is also a function of numerical aperture since gap must accompany angular misalignment. From a practical standpoint, there has to be some end separation. Having the ends of the fibers just touching would be most desirable, but joining and separating the connector halves very many times would soon result in some end damage to the fibers. To preclude scattering losses at the interconnection, fiber ends must be finished optically smooth.

Each of the foregoing losses, both intrinsic and extrinsic, assume optical energy is uniformly distributed over all modes. Each is independent of the others. Therefore, all losses must be added to calculate the total fiber optic link losses, that will include:

- Launching efficiency of the emitter

- Fiber attenuation against length

- Joint or connector losses

- Coupling efficiency of the detector

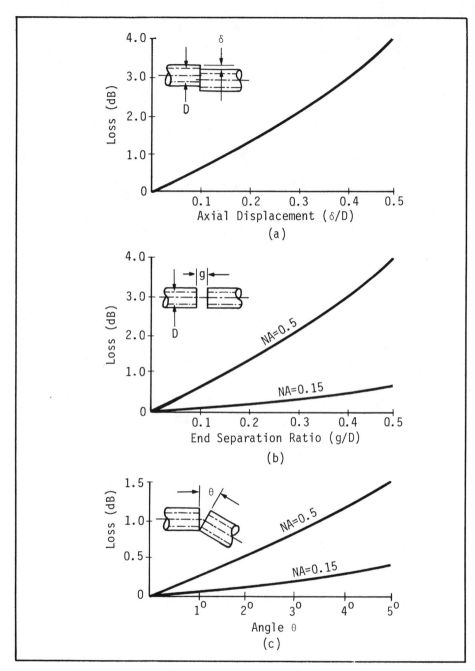

Figure 4.11 - Losses in Fiber Optic Connections.

4.5.3 Fiber Optic Couplers [25, 26]

Fiber optics are suitable for data bus distribution, owing to high bandwidth capacity, and their non-metallic nature. The fiber optic bus does not take on the same architectural configuration as the coaxial bus due to the add/drop coupler elements and their associated connectors. Instead, a special class of couplers is needed to perform signal distribution in bus (multiterminal) data communications systems. There are two popular types of optical bus couplers, the *tee* and the *star* couplers.

A *tee*-type optical coupler, which is compatible with a variety of fiber optic cable interconnections, is shown in Fig. 4.12a. Tee couplers provide at each terminal a technique to inject or remove a signal from a fiber optic trunk line. For multi-terminal networks, the serial distribution scheme using tee couplers provides advantages of flexibility in the number and location of the distribution paths or drops and minimizes the amount of fiber used in comparison with a star system.

The *star* element (Fig. 4.12b) is a simple optical splitting element that divides the input optical power from one input fiber into many output fibers or vice versa. In a star system, only one terminal at a given time in a given transmission band may broadcast. Time Division Multiple Access (TDMA) data transmission formats are usually used in these systems.

Work in this area includes a recently introduced liquid crystal device that efficiently switches signals between multimode optical fibers operating in the 0.4 to 2.5 μm wavelength region.[27] In general, a hybrid approach to the fiber optic bus would be implemented where optical splitters or mixed elements and add/drop elements are used together.

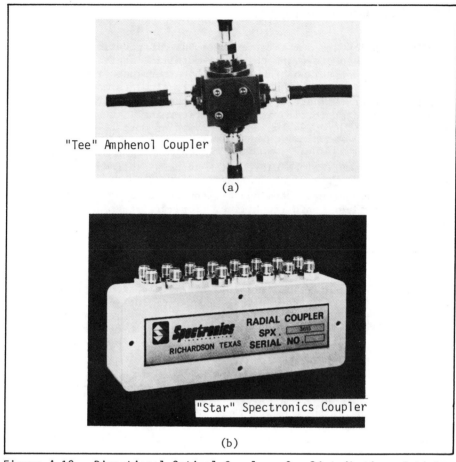

"Tee" Amphenol Coupler

(a)

"Star" Spectronics Coupler

(b)

Figure 4.12 - Directional Optical Couplers for Distribution of Light.

4.6 FIBER OPTIC DATA TRANSMISSION LINK

The basic elements of a fiber optic link system are shown in Fig. 4.13.[28] In general, all systems would follow the configuration shown in the block diagram. If there are any differences, they would be primarily in the elements that are indicated by the broken lines.[8] In the case of the simplest system, the active connector concept, the broken lines simply represent through connections. The transmitter would have only a light emitting source. There would be no repeater, and the receiver would only have a photodetector. The next stage of complexity might be a system using direct linear modulation of the sources with an amplifier included in the transmitter, the receiver, and the repeater (when required).

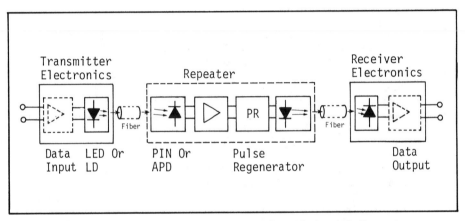

Figure 4.13 - Typical Optical Communication Link with Repeater.

When a laser source is used and the system is non-repeated, a relatively simple encoding method, such as pulse position modulation, would be adequate. For trunk communications, however, where a link might include several repeaters, a more complex encoding would be used. In this case, the repeater would contain circuits for the amplification, reshaping and retiming of signals.

The two basic types of optical fiber links, digital and analog, are discussed in the two subsequent chapters.

4.7 REFERENCES

1. Batchelder, G.R.; *The Communications Revolution 1976 -( )*; National News Conference St. Regis Hotel, New York, 27 April 1976.

2. Hudson, M.C., and Dobson, P.J.; *Fiberoptic Cable Technology*; Microwave Journal, pp. 46-53, July 1979.

3. Bowen, T. and Schumacher, W.; *Fiber Optic Connector Developments; Moving to Annul Coupling Mismatches;* Microwave Journal, pp. 55-59, July 1979.

4. Giallorenzi, T.G.; *Optical Communications Research and Technology: Fiber Optics;* Proceedings of the IEEE, Vol. 66, No. 7, pp. 744-780, July 1978.

5. Conradi, J., Kapron, F.P. and Dyment, J.C.; *Fiber-Optical Transmission Between 0.8 and 1.4 μm;* IEEE Journal of Solid-State Circuits, Vol. SC-13, No. 1, pp. 106-119, February 1978.

6. Amphenol FOC Series; *Amphenol Fiber Optic Interconnections;* Bunker Ramo, Amphenol North America, Danbury, Ct.

7. Margolin, B.; *Fiber Optic Links Come of Age;* Electronic Products Magazine, pp. 40-43, November 1979.

8. Sandbank, C.P.; *Fiber Optic Communications: A Survey;* Electrical Communication, Vol. 50, No. 1, pp. 20-27, 1975.

9. Ramsay, M.M., Hockham, G.A. and Kao, K.C.; *Propagation in Optical Fiber Waveguides;* Electrical Communication, Vol. 50, No. 3, pp. 162-169, 1975.

10. Marcatili, E.A.J.; *Modal Dispersion in Optical Fibers with Arbitrary Numerical Aperture and Profile Dispersion;* Bell System Technical Journal, Vol. 56, pp. 49-63, Jan. 1977.

11. Knoblauch, G.; *Fiber Optic Communications in Industry;* Siemens Components, Vol. XV, No. 3, pp. 144-150, 1980.

12. Monthierth, M.R.: *Optical First Drawing Techniques;* Optical Spectra, Oct. 1978.

13. Simpson, J.R., et al; *Preform Fabrication at High Deposition Rates Using MCVD;* 82nd Annual Meeting and Exposition, American Ceramic Society, Chicago, Ill., April 27-30, 1980.

14. Schultz, P.C.; *Progress on Optical Waveguide Fabrication*

Processes; 82nd Annual Meeting and Exposition, American Ceramic Society, Chicago, Ill., April 27–30, 1980.

15. Bark, P.R., et al; *Fiber Optic Cable Design, Testing and Installation Experiences;* 1980 International Wire and Cable Symposium, November, 1980.

16. Kleekamp, C. and Metcalf, Bruce; *Designer's Guide to Fiber Optics-Part I;* EDN, January 5, 1978, and Part 4, March 5, 1978.

17. Ellis, J. and Massanari, G,; *Current Trends in Optic Cable Construction;* Electro Optic Laser '80 Boston, Ma., November 21, 1980.

18. Elphick, M.; *Rapid Improvements in Fibers, Connectors, Emitters and Detectors Promise Wide Use Soon;* High Technology, pp. 59–63, April 1980.

19. Kessler, J.N.; *Fiber-Optic Connectors: Prices Drop, Performance Rises;* Electro-Optical Systems Design, pp. 29–33, October 1979.

20. Null, G.M., Uradnisheck, J. and McCartney, R.L.; *Three Technologies Forge a Better Fiber-Optic Link;* Electronic Design, 1980.

21. Kohanzadeh, Y.; *Hot Splices of Optical Waveguide Fibers;* Applied Optics, Vol. 15, pp. 783–795, March 1976.

22. Kendrick, A.; *Fusion Splicing: Key to Low-Loss Fiber Optics;* Design Engineering, March 1979.

23. Borsuk, L.M.; *What you Should Know About Fiber Optics;* Digital Design, pp. 56–68, August 1978.

24. Bowen, T. and Gempe, H.; *Impact of Coupling Efficiency on Fiber Optic System Performance;* Electro-Optical Systems Design, pp. 35–43, Aug. 1980.

25. Hudspeth, W.S.; *Fiber Optic Connectors-Still a Budding Technology;* Electro-Optical Systems, pp. 46–49, October 1978.

26. Ormond, T.; *Fiber-Optic Components;* EDN, pp. 87–96, March 20, 1979.

27. Zeskind, D.; *Liquid-Crystal Fiber-Optic Switch Promises Wide Data-comm Use;* EDN, Vol. 24, No. 12, pp. 57–60, 1979.

28. Baues, P.; *The Anatomy of a Fiber Optic Link;* Control Engineering, pp. 46–49, August 1979.

CHAPTER 5

DIGITAL FIBER OPTIC LINKS

Various fiber optic components can be combined to build digital links of various lengths using general-purpose interface ICs. To facilitate interfacing in digital data communications applications, the link transmitter and receiver usually employ standard TTL devices. This chapter focuses on the digital link's functional circuits and examines some of their design characteristics and applications.

5.1 INTRODUCTION

Increased performance requirements plus availability of high quality fiber optic components has created the need for easy-to-use digital electronic circuits to interface with optoelectronic components.[1,2] These components are fiber optic line drivers and receivers compatible with the optical signals with a variety of optical transducers, i.e., LEDs, APDs, etc., electrically compatible with standard logic families.[3]

All digital data links consist of a transmitter that supplies power to modulate a fast rise-time, solid-state light source, a source-to-fiber interface, a suitable length of optical fiber, a fiber-to-detector interface, and finally an optical receiver.[4] Digital data link performance depends greatly on how carefully one handles the optical components during the circuit construction.

Even though optical fiber cables provide unsurpassed immunity over their length to RFI, EMI, and EMP compared to conventional wires, the digital modules themselves are not designed specifically to provide high levels of immunity to electromagnetic effects. Therefore, in such environments, the transmitter, receiver, and associated data processing equipment should be mounted in shielded enclosures. For these installations, the optical fiber cable should pass to the external environment through a metal tube which can eliminate extraneous interference if properly designed.[5,6]

5.2 COMPONENT CONFIGURATION IN A DIGITAL FIBER OPTIC LINK

The component configuration in a digital fiber optic system is shown in Fig. 5.1. The signals are first amplified and then transformed into optical signals in an input optocoupler. The input amplifier has the task of interfacing the electrical signals (current amplification) to the parameters of the optocoupler. The optical cable has connectors at both ends and can be of any of the available types, depending on the attenuation which can be afforded.

At the output of the optical cable, the light is converted back into electrical signals (photocurrent) in an input optocoupler and the resulting photocurrent is finally amplified and processed in the output amplifier.

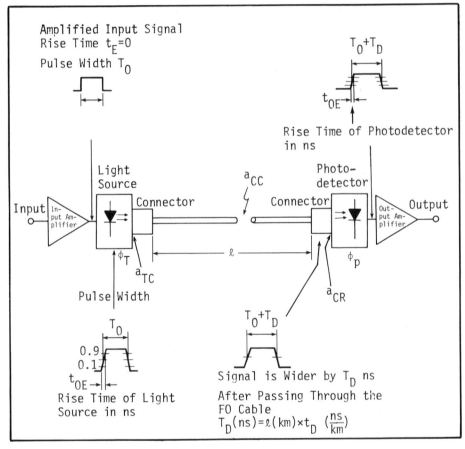

Figure 5.1 - Digital Fiber Optic Transmission System.

5.2.1 Form of Input and Output Signals

In the standard ITT range, the input to the transmitter is applied via a standard interface, normally TTL, so that the user may clearly understand what he needs to do in the external circuits to drive the transmitter. The convention used is that a TTL high turns the emitter *on*, whereas a TTL low turns it *off*. This format is suitable for non-return to zero (NRZ) codes provided that an LED or other continuous emitter is used in the output of the transmitter circuit, and designs permit operation up to 20 Mb/s. Where long-haul applications are contemplated, it sometimes becomes necessary to resort to laser output transmitters.[2]

The receiver consists of a photodetector upon which the incoming light signal impinges, coupled to an amplifier, which is followed by a decision-making comparator, giving TTL high or low to the output depending on whether light is received by the photodetector or not. The photodetector can be either an avalanche photodiode (APD) or a PIN junction diode, and can be either DC coupled to the amplifier or AC coupled.

Transmitter and receiver units for optical cable links in information systems can be separately plated on PC cards for rack mounting, or can be developed into integrated versions using existing ICs and placed on the same PC board[1,7,8] as shown in Fig. 5.2.

5.2.2 Flux Budgeting

In order to establish the flux requirements for a fiber optic system, the characteristics of the receiver noise and bandwidth, coupling losses at connectors, and transmission loss in the cable should be taken into consideration.

In Fig. 5.1 the flux which the transmitter must produce is determined from the expression:[9]

$$10 \log \frac{\phi_T}{\phi_R} = a_0 \ell + a_{TC} + a_{CR} + n a_{CC} + a_M \qquad (5.1)$$

where,

ϕ_T = Flux (in μW) available from the transmitter

ϕ_R = Flux (in μW) required by the receiver

a_0 = Fiber attenuation constant (dB/Km)

a_{TC} = Transmitter-to-fiber coupling loss (dB)

ℓ = Fiber length (Km)

a_{CC} = Fiber-to-Fiber loss for in-line connectors (dB)

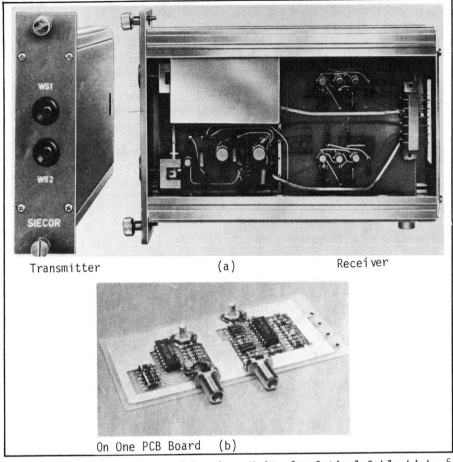

Transmitter   (a)   Receiver

On One PCB Board   (b)

Figure 5.2 - Transmitter and Receiver Units for Optical Cable Links.[6]

n = Number of in-line connectors (excluding the end connectors)

a_{CR} = Fiber-to-receiver coupling loss (dB)

a_M = Margin (dB), chosen by the designer, by which the
transmitter flux exceeds the system requirement.

For flux measurements, either a high-speed photodetector and
oscilloscope could be used to measure the excursion flux, or an aver-
age-reading flux meter to measure $\Delta\phi$.

5.2.3 Pulse Spreading and Rise Times Along the Optical Path [10]

This description deals only with that part of the rise time introduced by the components of the optical-fiber cable routes, excluding input and output amplifiers (Fig. 5.1). A rise time of $t_E = 0$ is assumed for the amplified input signal. The rise time is the time required for a pulse to rise from 10% to 90% of its peak value and the pulse width is measured at half the peak amplitude. The total rise time, t_r, of the optical fiber route (excluding input and output) is:

$$t_r = \sqrt{t_{EO}^2 + t_{OE}^2} \tag{5.2}$$

where, t_{EO} = Rise time of optical source

t_{OE} = Rise time of photodetector

On the other hand, the pulse width, T, at the output of the photodetector is given by:

$$T = T_0 + T_D = T_0 + \ell \times t_D [ns] \tag{5.3}$$

where t_D is the pulsespread (ns/Km) on the optical fiber cable given in the data sheets for specified optical wavelength, $\lambda + \Delta\lambda/2$ nm, ($\Delta\lambda$ is the spectral bandwidth of a light source, electro-optical transducer). The value of t_D increases with increasing spectral bandwidth $\Delta\lambda$.

5.3 TRANSMITTER NETWORKS

Sources for fiber-optic systems include solid-state, high-radiance LEDs and injection diode lasers (IDLs). Once chosen, the appropriate source must be designed into a complete transmitter circuit. The important parameters of the transmitter include the output-power level, on/off or extinction ratio, switching speed and spectral of the optical signal.

5.3.1 Fiber Optic Transmitter Block Diagrams

A digital transmitter network block diagram is shown in Fig. 5.3. When the source is an LED, the driver element can be as simple as the one shown in Figure 5.4. This element can be an IC driver, inverter, or gate with the inputs tied together, such that it can supply adequate drive for the LED. The GaAlAs infrared-emitting diode, for an on-condition of 100 mA forward current, has a typical forward voltage of 1.6V. Junction capacitance at zero volts is typically 120 pF, and diode series resistance is in the range of 1.5 to 6 Ω. Since the radiant power output is proportional to the input drive current, the ideal drive circuit would be a current source.

A laser transmitter is somewhat more complicated than an LED transmitter, since the laser itself is a threshold device in which the threshold changes with temperature and aging. There is a trade-off between laser switching speed and the *on*-to-*off* ratio (extinction ratio) of the laser light pulses. The tradeoff establishes the laser dc bias. Setting the bias current above the threshold maximizes switching speed, but results in a nonoptimal extinction ratio. The extinction ratio is increased by setting the dc bias below threshold at the expense of reduced switching speed. The laser driver dc-biases the laser slightly below threshold and adds enough signal current to bring the laser to the *on* state. Figure 5.5 shows a block diagram of a laser-driven circuit in which two factors (at the summing point) influence the modulation.[11] One factor is the optical feedback and the other is the bias-current reference. As the laser threshold changes, the feedback circuitry adjusts the bias current to stabilize the average light output of the laser. Figure 5.5 shows one type of feedback circuit in which a 5% optical tap monitors the average laser output power. As an alternative to the optical tap method, light from the back facet of the laser can be used if the front and back outputs track one another over temperature and time.

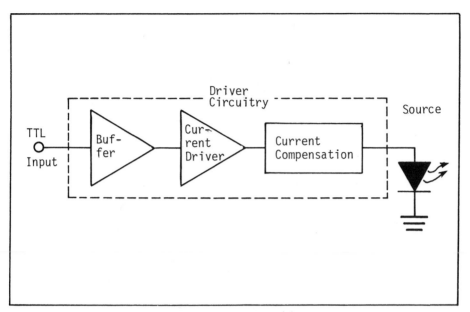

Figure 5.3 - Digital Transmitter Block Diagram.

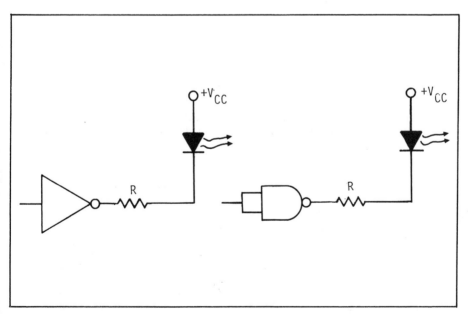

Figure 5.4 - Basic Digital Drivers for LEDs. The Resistors are for
Current Limiting and Should be Chosen Accordingly.

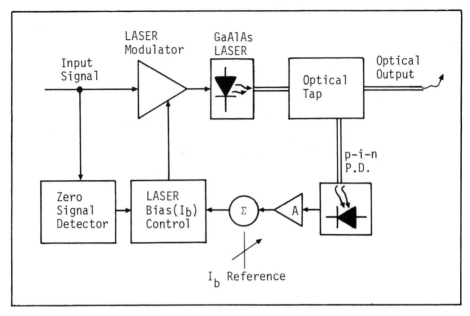

Figure 5.5 - Block Diagram of Laser Driven Circuit

5.3.2 Driver Modulating Schemes

The circuits in Fig. 5.4 will only reproduce what is fed to them
on a one-to-one basis. To pulse modulate the LED as a function of
analog input signals, however, other circuitry is needed. There are
several methods of pulse modulation, with three of the more common
illustrated in Fig. 5.6.

In the first scheme, the number of pulses per unit time is
varied at the input signal rate. This system is essentially simple
FM, In the second, the width of a fixed series of pulses is varied
at the information rate, while in the third, the position of the
pulse width with respect to a reference pulse is varied.

Figure 5.7 shows a very simple uni-junction pulse modulator
which achieves pulse rate modulation. The circuit operates at a
repetition rate determined by the values of R and C and the point at
which the PNP transistor is biased. Typical center frequencies for
this type of set-up are 5-15 kHz.[4] Incoming analog signals, suitably
offset to allow for transistor biasing, vary the conduction of the

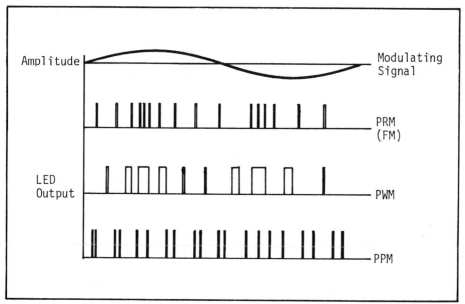

Figure 5.6 - Three Common Pulse Modulating Schemes

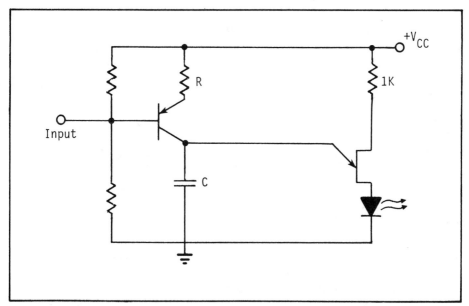

Figure 5.7 - Simple UJT PRM Modulator.

transistor and, consequently, the operating frequency of the UJT oscillator.

5.3.3 Design Examples of Transmitting Circuits

A transmitter circuit that requires only a single 5 V supply is shown in Fig. 5.8.[12] All inputs and outputs function at TTL levels. Besides the logic arrangement, waveforms for signal currents I_A and I_B and the resulting waveforms for the output flux are also shown.

In Fig. 5.8, five important technical design characteristics are highlighted:

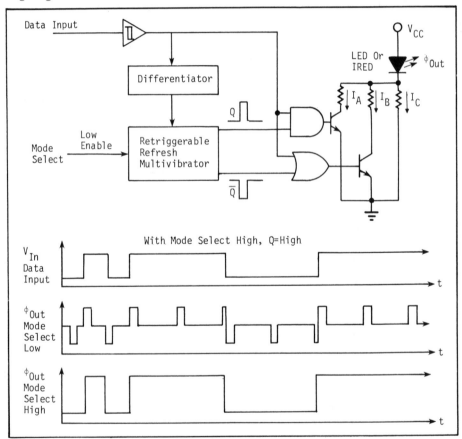

Figure 5.8 - Transmitter Network with Single 5V Supply. In Addition to 5V and Ground, There are Two TTL Level Inputs: Data Input and Mode Select *(After Reference 12)*.

- The bias current I_C is never turned off, even when the transmitter is operated in the externally-coded mode (mode select high). This is done to enhance switching speed of the LED in either the internally or externally coded mode. The bias current also stabilizes the flux excursion ratio (k) symmetry in the internally coded mode.

- The current I_C produces low-level flux ϕ_L. Mid-level flux ϕ_M requires I_B+I_C, while high-level flux requires $I_A+I_B+I_C$. It must be emphasized here that for the receiver, the excursion flux $(\Delta\phi)$, produced by switching I_A and I_B, is the most important parameter of the transmitter. Average flux is, of course, related to the excursion flux, but is not as important in establishing the S/N ratio of the system.

- With mode select low and a typical 500-kHz signal at data input, there will be only one refresh pulse generated in each logic state. Excursions $(\phi_H - \phi_M)$ and $(\phi_M-\phi_L)$ are nearly balanced. Thus, an average reading flux meter will indicate the mid-level flux (ϕ_m) within $\pm 0.-6\%$, depending on whether the flux excursion ratio is at its maximum or minimum limit, respectively.

- With mode select low, any data input transmission (either high-low or low-high) retriggers the refresh multivibrator to start a new train of pulses. All refresh pulses for either logic state have the same duration. This keeps average flux near mid level, even when the duration in either logic state of arbitrarily timed input data is short. Any fresh pulse is overridden by the occurence of a data input transition. Thus, there is no additional jitter when the duration of the data input in either state is at or near the same length of time as the refresh interval. The refresh interval is long, relative to the refresh pulse duration, making a duty factor of approximately 2%. This also is done to keep the average flux near mid-level regardless of how long data input remains in either logic state. The only condition under which the average flux can deviate significantly from mid-level occurs when data remains in one state for a period of time less than the duration of the refresh pulse. If this is likely to occur, the format should be configured so that the numbers of 1's and 0's are balanced as they would be in Manchester code. Observing this data format allows the use of the internally coded mode of this fiber optic system at data rates ranging from arbitrarily low-to-higher than 10M baud, the absolute limit being that at which signal intervals become as short as t_{PHL} (propagation delay for high-to-low transistion) and/or t_{PLH} (propagation delay for low-to-high transition).

- With mode select high, the $\overline{Q}$ output of the refresh multivibrator is high (and $\overline{Q}$ is low). In this condition, I_A and I_B are both ON when data input is high, and both are OFF when it is low. This makes the output flux excursion a logical replica of the input data.

Another representative driving network using ECL circuitry is shown in Fig. 5.9. The driver switches 50 mA *on* and *off* with rise and fall times under 5 ns.[6]

The LED in this circuit, a Motorola MFOE103F FOAC, typically switches in 20 ns and launches 35 μW of optical power, centered about 910 nm, into its fiber.

For faster links, the TXES-491 LED from TI switches with 6 ns typical rise or fall times at 850 nm. At 50 mA drive, this LED launches about the same optical power into a 0.2 mm fiber as the Motorola LED. For longer distance links, the SE3352-03 LED from Spectronics typically launches 200 μW into a 0.2 mm fiber at a 50 mA drive. Spectronics' LED switches somewhat slower than TI's - 12 ns at 820 nm. The optical output for all three LEDs approximately doubles when the drive is doubled to 100 mA.

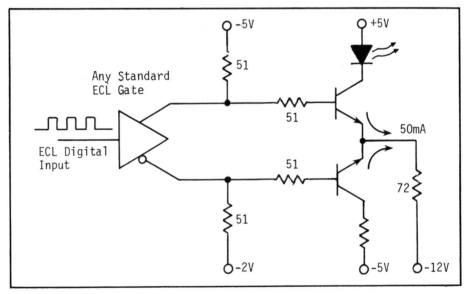

Figure 5.9 - ECL Driven Network.

5.4 RECEIVER NETWORKS

The design of receivers for most telecommunications applications is largely dominated by the performance of the photodetector, but for other applications such as asynchronous data systems, where dc coupling is required, the preamplifier also becomes very important.

5.4.1 General Requirements

The frequency of the data rate desired primarily determines the choice of photodetector to be used. For long fiber links, the noise in the system also dictates the choice of a detector. The design objective is to maximize the signal-to-noise ratio (S/N) at the output of the preamplifier, i.e., immediately prior to a binary decision being made in the regenerator. This will set the minimum required received signal power for a given system error-rate performance which could typically be 1 in 10^9. The very steep dependence of error rate on S/N is shown in Fig. 5.10, from which it can be seen that 0.5-dB improvement in S/N can decrease the error rate by an order of magnitude.[13]

The lowest cost detectors are phototransistors, which offer high gain, but suffer with rise and fall times on the order of a few microseconds. This limits their usefulness to low rates and, as a second consideration, they are also nonlinear. While this is not too serious for pulse receiving, it does cause problems with analog signals.[4] Low-cost field effect phototransistors are also available. These may be used where high gain is required from the detector itself. They are much faster than bipolar phototransistors and offer rise and fall times on the order of 30-150 nanoseconds. This makes them suitable for low-to-medium speed applications, but not adequate for high speed work. For medium-to-high speed applications, PIN photodiodes will be required. These devices offer rise and fall times in the nanosecond and subnanosecond area as well as a linear response. Typical frequency response is to 1 GHz.

A photodiode's most fundamental characteristic is its responsivity, i.e., the amount of current it will produce in response to the incident light power. In applications requiring very high sensitivity, avalanche photodiodes (APDs) must be used. The avalanche photodiode requires 150-300 volts of reverse bias to operate. The gain-bandwidth product of an avalanche photodiode is in the neighborhood of 100 GHz. The drawbacks to avalanche photodiodes are the high bias voltage needed, and the temperature compensation necessary for stable operation.

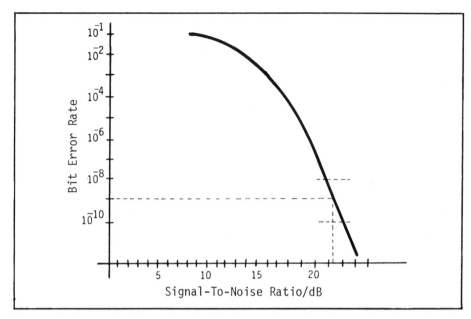

Figure 5.10 - Relationship of Bit Error Rate to Signal-To-Noise Ratio.

5.4.2 Design Considerations and Noise Problems

The most challenging aspect of many fiber optic links is the design of the receiver. The receiver must convert the low-level current output of a photodiode to a high-level analog or digital signal with accuracy and speed. As was mentioned previously, the design objective is to maximize the signal-to-noise ratio (S/N) at the output of the preamplifier.

The receiver (Fig. 5.11) in a fiber optic link is more complex than the transmitter, as a block-by-block examination reveals.[14] Because the detector is a high-impedance source with a tiny small-signal output, it is difficult to interface without introducing noise, RFI and reactive loads, which degrade signal quality. For this reason, the receiver's current-to-voltage converter (CVC) usually couples as closely as possible to the detector, with the interface between them often shielded from outside interference.

The linear voltage amplifier, the third element in the receiver's block diagram, should have gain sufficient to amplify the CVC's expected noise nearly up to the minimum threshold level of the fifth block, the amplitude detector. Any gain greater than the minimum threshold level merely amplifies both noise and signal and does

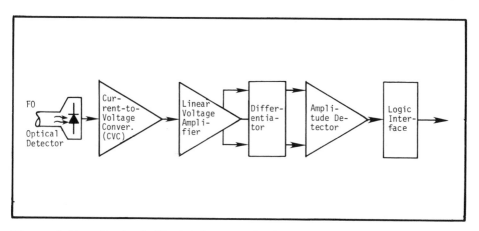

Figure 5.11 - Typical Block Diagram of Receiver in a FO Link.

nothing more than increase the amplitude detector's required threshold
level. The voltage amplifier's bandwidth and rise time are also
critical.

The rigorous noise analysis of an optical receiver is very complex.[15,16] To simplify things, the signal and noise current contributions are taken as shown by the current generators in Fig. 5.12.
The signal current, I_P, due to incident radiation, P_L, is given by:

$$I_P = \eta \frac{e}{h\nu} P_L \quad , \tag{5.4}$$

where e is the electronic charge, h is planck's constant, ν is the
optical frequency, and η (quantum efficiency) is the ratio of free
carriers to photons.

Due to the random nature of photon arrival, the signal will have
shot noise current associated with it. Thus:

$$\sqrt{\overline{i_S^2}} = \sqrt{2eI_P B} \quad , \tag{5.5}$$

where B is the modulation bandwidth.

If the photodiode is operated in an avalanche mode, a further
noise term is included to take account of the statistical nature of
the avalanche gain (M). That is:

$$\sqrt{\overline{i_\eta^2}} = M\sqrt{\overline{i_S^2}(M^X - 1)} \quad , \tag{5.6}$$

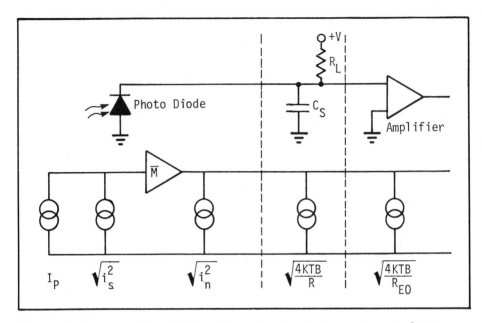

Figure 5.12 - Signal and Noise Currents in Simplified Equivalent Circuit of Receiver Input Components.

where x is known as the excess noise factor. If M=1, as with PIN diodes, the excess noise term vanishes. The load resistor and amplifier contribute thermal noise as shown.

From these equations it can be concluded that the diode noise is signal-dependent and, due to the excess noise factor, increases faster than the signal current as M is increased. Thus, if M is very large, the diode is the dominant noise source; whereas if D is very small, the load resistor and amplifier noise contributions dominate. An intermediate value of M exists which gives the best signal-to-noise performance at a given signal level, and this is the value the design procedure attempts to optimize. In the case of a PIN diode receiver, the largest value of diode load resistance, R, gives minimum noise, but the resulting bandwidth due to the time constant at this point must be taken into account. This usually puts an upper bound on the highest practical value of load resistor that can be tolerated. If the receiver sensitivity is defined as the minimum optical power to give a specified S/N ratio, then an APD can offer up to 20-dB advantage in sensitivity over a PIN diode, unless exceptional care is taken to minimize the stray capacitance, C. This capacitance arises from the photodiode junction capacitance, capacitance in the diode package and the input capacitance of the preamplifier. Choice of diode can minimize the first cause while particular attention to constructional techniques and circuit design can reduce the effects of the second and third.

Returning to Fig. 5.11, and assuming that the transmitter and fiber rise times are so short and their contributions to system rise time are negliglible, it is possible to write the relationship:[14]

$$t_{RS} = \sqrt{(t_{RIDP})^2 + (t_{RVA})^2} , \tag{5.7}$$

where t_{RS} is the desired system rise time at the voltage amplifier's output, t_{RIDP} is the rise time of the integrated optical decoded/preamplifier (IDP) and t_{RVA} is the voltage amplifier's required rise time. In order to ensure that the voltage amplifier does not degrade system rise time by more than 10%, the following relationship must hold:

$$\sqrt{(t_{RIDP})^2 + (t_{RVA})^2} \leq 1.1 t_{RIDP} \tag{5.8}$$

Hence,
$$t_{RVA} \leq 0.45 t_{RIDP} \tag{5.9}$$

Unfortunately, other considerations preclude making the amplifier's rise time too short. As a rule, the linear amplifier's bandwidth should be limited so that its noise contribution is about 50% of the IDP noise. Amplitude-detector threshold level is then a function of IDP noise level, which is a desirable condition.

Special attention must be given to the R and C values. An edge-coupled receiver strips off duty-cycle-related baseline variations from the data stream. Inserted between the linear voltage amplifier and the amplitude detector, the differentiator performs this function, and the choice of R and C values for it requires careful consideration.

As a final block in the receiver, the logic interface serves to generate a standard logic level and provide sufficient drive capability to simplify interfacing. The amplitude detector actually generates the receiver's logic level. However, the interface block buffers the detector's output and provides some isolation from the outside world. Additionally, the interface may include an emitter-follower in order to provide the necessary drive to external circuitry.

As far as external noise is concerned, even though optical fiber cables provide unsurpassed immunity over their length to RFI, EMI, and EMP compared to conventional wires, the digital modules

themselves are not designed specifically to provide high levels of
immunity to electromagnetic effects. Therefore, in such environments,
the transmitter, receiver, and associated data processing equipment
should be mounted in shielded enclosures.[5,6] For these installations,
the optical fiber cable should pass to the external environment
through a metal tube which can eliminate extraneous interference if
properly designed.

5.4.3 Representative Receiving Circuits

Depending on the particular applications, a fiber optic receiving circuit may have all or some of the components of the block diagram of Fig. 5.11. Following is a discussion of three representative
receiving circuits.

Figure 5.13 shows the electrical diagram of Spectronics' SD-4323-
002 detector which is integrated on the same chip with a transistor
amplifier.[17] The responsivity of the entire device is approximately
30 to 40 $\mu A/\mu W$ for receiving a wavelength of 820 nm from a 100 μm
core cable and driving a 5V, 200 Ω load. The device is fast enough
to produce a 2-MHz analog or 1-Mbit/s digital output. The dark current is less than 1μA at 5.5 V, so that *on* and *off* states are easy to
distinguish, and the device interfaces readily with TTL or comparator
circuitry.

The high-speed SD-4478 photodiode has a responsivity of 0.45 $\mu A/$
μW at 820 nm, and a peak responsivity of approximately 0.5 $\mu A/\mu W$ at
907 nm. Because its rise time is less than 5 ns with 15-V bias voltage, it supports 50-Mbits/s transmission. The rise time can be
further shortened by increasing the bias voltage. With 50-V bias,
for example, T_r drops to 2 to 3 ns. Even with 5-V bias and reduced
responsivity, the device is fast enough to accommodate 2 and 3-Mbit/s
transmissions.

The LH0082 fiber optic receiver produced by National Semiconductor contains a preamplifier with a gain-bandwidth product of nearly
3 GHz. The LH0082 will provide the sensitivity and speed necessary
for the example application, and it also includes a comparator for
providing a TTL/DTL/CMOS compatible output. Figure 5.14 shows how to
use the LH0082 as a 6-Mbits/s, 300 nW sensitivity fiber optic re -
ceiver.[17,18] The only external components needed are the photodiode,
a power supply decoupling resistor and two bypass capacitors. With
a high sensitivity of 30 nW, data rates range from dc to only 1 Mbit/s.

A receiver network that preserves the desirable characteristics
of both ac-and dc- coupled circuits is shown in Fig. 5.15.[20] By
differentiating the input waveform, the circuit produces a positive
pulse on the positive going logic transition and a negative pulse on

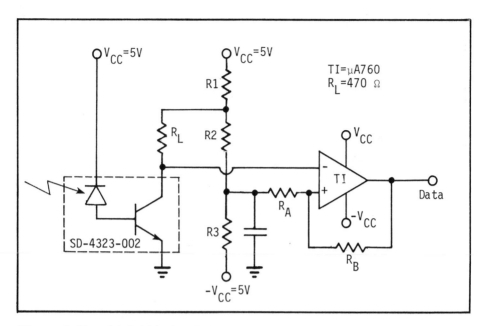

Figure 5.13 - DC-1 Mbit/s Digital Receiver.

the negative going transition. Standard ac-coupled amplifier techniques provide amplification. To produce the original pulse train, the
positive pulse sets an RS flip-flop, and the negative pulse resets it.
Thus, the receiver reconstructs the pulse train for any pulse sequence,
just as a dc receiver does, and provides the ease of operation of an
ac-coupled version.

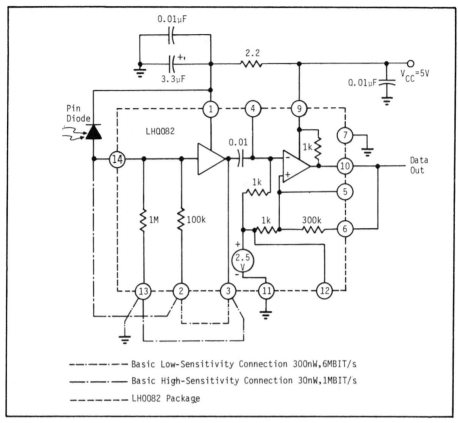

Figure 5.14 - A 6 Mbits/s Fiber Optic Receiver Built Around the LH0082 National Semiconductor Device.

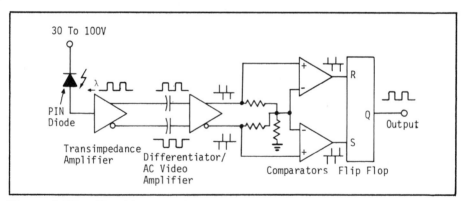

Figure 5.15 - The Circuit Differentiates its Input to Produce Positive and Negative Pulses.

5.5 REPEATER NETWORKS

The capacity of a transmission system helps determine whether it is practical. Another factor is repeater spacing, i.e., how long segments of cable can be before the signal must be regenerated. This spacing depends on the power of the signal as it travels through the system. The signal that arrives at a repeater or receiver must be strong enough to trigger the acknowledgement of pulses in the correct time slot. If so much signal power were lost that the signal had to be regenerated every few hundred meters, the system would not be practical.[21]

5.5.1 Components of Repeater Networks

Each repeater consists of a line receiver, an amplifier and a line transmitter. In addition, it may perform regeneration and re-timing for pulse code modulation (PCM) signals, and also provide supervision, alarm and other functions.[22] For duplex operation, the repeaters have two sets of receiver and transmitter circuits. In a simplex arrangement the repeaters have only one set of receivers and transmitters as shown in the block diagram of Fig. 5.16.[23]

In this network the avalanche photodiode (APD) converts the optical signal into a low-level electrical signal. A transimpedance front end and preceding analog amplifiers are used to boost the signal. Filters are used to help eliminate low-frequency noise and yet not restrict data flow. The amplitude of the resulting signal is monitored by a slow response automatic gain control (AGC) processing circuit that adjusts the amplifier gain and APD voltage to accommodate high-level signals. After amplification of the signal, a balanced clamp and comparator help regenerate the risetime of pulse edges even though there is no timing recovery to regenerate pulse edge position and pulse width. The driver circuit further boosts the digital signal to drive the semiconductor laser. The optical power emitter from one facet of the laser is coupled into the fiber. The optical output of the laser is stabilized by capturing the optical power emitted from the other facet of the laser, measuring the average power, and using the resulting electrical signal in a feedback loop to control the laser drive current.[23]

5.5.2 Repeater Spacing

The spacing of repeaters in fiber optic systems depends on the extent of signal dispersion and on the strength of the signal. At lower rates, signal loss is limiting, while at higher bit rates,

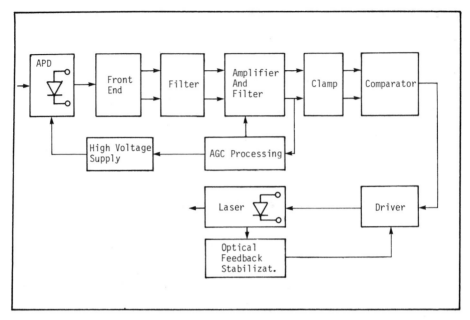

Figure 5.16 - Block Diagram of Repeater Network.

dispersion is limiting. An LED couples a weaker signal into the
fiber than does a laser, and chromatic dispersion then limits the
pulse rate. Laser transmission is so little affected by chromatic
dispersion that modal dispersion is the limiting factor.[21] These
limitations and other details are shown in Fig. 5.17.[24] As can be
seen, the repeater spacing for the different bit rates is always
higher than the corresponding spacing in the copper cable systems.
This fact plays an important role in the economy of the fiber systems with respect to the wire systems. Table 5.1 shows that previous
coaxial systems, for example, required repeaters at 1 to 2Km intervals.[25] Today's production fiber systems space these by 8 to 10Km,
while future systems now under development will be able to increase
that to 40 or even to 100 Km, if required.

5.5.3 Repeater Powering for Fiber Optic Communication
Systems [26]

A major problem in fiber optic systems is that of repeater powering, since the optical fiber cannot serve as a powering medium as
copper cables can. In a fiber optic system, a number of repeater
chains have to share a single powering pair for the sake of economy
to increase I_R in Fig. 2.18a. If conventional repeater terminal
voltage E_R (approximately 10 V) has to be maintained, quite a large

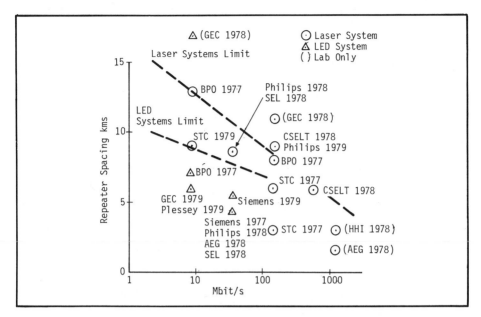

Figure 5.17 - Repeater Spacing of Trial Systems from 1977 to 1980 (*Adapted from Reference 24*).

powering voltage E_T would be required. The parallel powering plan depicted in Fig. 2.18b is superior to the series powering method, as can be seen from Fig. 5.18c. The comparison of powering plans for cases using ordinary dc analysis is quite cumbersome. A method that significantly simplifies the analysis has been proposed in Reference 26.

Japanese researchers, using the right wavelength (1.27 µm), specially constructed grated-index optical fibers, and a low-pass fusion splicing method, managed to transmit 32 Mbits/s pulse-code modulated (PCM) signals over a 53.5-km stretch without intermediate repeaters.[27] The total loss of the 53.3-km system was only about 35-dB. This suggests several practical advantages for installing fiber-optic systems over established routes. For example, repeaters could be located in established telephone company offices, not in manholes, and this fact would ease power-supply requirements, and reduce the cost, effort and time of maintenance operations.

Table 5.1 - Comparison of Fiber Optic and Conventional Systems *

RATE	VF CIRCUIT CAPACITY	CONVENTIONAL TECHNOLOGY	REPEATER SPACING	OPTICAL FIBER TECHNOLOGY REPEATER SPACING	
				6 dB/km	3.5 dB/km
T1 1.544 Mbps	24	Twisted Wire Pair 2 GHz Digital Radio	1.8 km 20-32 km	9 km	15.4 km
T1 C 3.152 Mbps	48	Twisted Wire Pair 2-6 GHz Digital Radio	1.8 km 20-32 km	8.5 km	15 km
T2 6.312 Mbps	96	LO-Cap Twisted Wire 2-6 GHz Digital Radio	4-5 km 20-32 km	8 km	13.7 km
T3 44.736 Mbps	672	6-11 GHz Digital Radio	20-32 km	6.4 km	11 km
2 x T3 Not Yet Standardized	1344	6-11 GHz Digital Radio	20-32 km	5.9 km	10.1 km
T4 274.176 Mbps	4032	Air Dielectric Coax T8 GHz Digital Radio	1.6 km 4.8 km	4 km	8 km

* Adapted from Reference 25

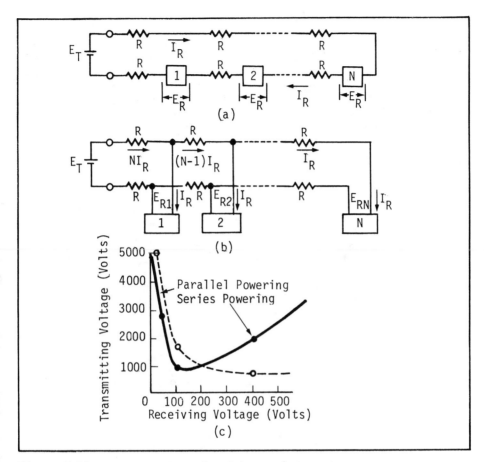

Figure 5.18 - Repeater Powering Schemes (*Adapted from Reference 26*).

5.6 FIBER OPTIC DATA BUS [28—32]

Conventional interface techniques in bus-organized systems, for high speed data transmission, suffer from pick-up disturbances, ground noise, delays due to reflections and the relatively high power dissipation due to required active or passive terminations at the end of the line. Fiber optics are well suited for data-bus applications offering many advantages over conventional wires.

5.6.1 General Features and Description

Data rates as high as 300 Mbytes/s are attainable with today's technology, and rates of 1 Gbyte/s are expected in the future. [28] This means that distributed processing systems will be set up so that 1000 terminals operating at 6 Mbytes/s can be connected over a single fiber-optic bus without repeaters.

In an optoelectronic data bus, the various stations are interconnected with flexible fiber optic bundles. The desired signal coupling device should provide the following functions: [29]

- A portion of the optical signal should be removed from the bus for detection.

- The undetected remainder of the optical signal should be passed on for distribution to the other terminals on the bus.

- The optical signal generated in that terminal should be coupled onto the bus and distributed to the other terminals.

Construction of a data bus requires the use of signal coupling devices which make it possible for each station to receive signals from the bus and transmit signals onto the bus. Fault isolation is necessary in an optoelectronic data bus system, so that failure on one of the stations on the bus will affect only that station and will leave the remainder of the bus unimpaired. In general, repeater systems are not employed in data buses because damage to one repeater would interrupt signal flow on the entire data bus.

5.6.2 Coupling Techniques

The couplers discussed in Chap. 4 add extra degrees of freedom to fiber optic systems designs. These couplers can add bidirectionality, signal mixing, and tapping to already well-known, single-optical-fiber advantages. Figure 5.19 shows systems which utilize reflective and transmissive star couplers. In these structures, signals from each source reach each detector.[30] The final important consideration in designing a multiuser communication system is the method by which data traffic is handled within the network. When numerous coupler components are chained together, a worst-case budget must be prepared to indicate the appropriate optical components.

5.6.3 Bus Structure Configurations

One special consideration required for data bus system architecture design is intermessage dynamic range. This is the difference in amplitude of signals from two different bus terminals. This situation occurs when bus transmission switches from a near (low-loss path) terminal to a distant (high-loss path) terminal. The receiving terminal must be able to detect, synchronize and receive both the highest and the lowest-amplitude signal. The lowest-amplitude signal is obtained from the worst-case power budget to each terminal. The highest amplitude signal is obtained from a best-case power budget to each terminal. The difference between these amplitudes is the intermessage dynamic range.[31]

Examples of optical data bus configurations are shown in Fig. 5.20.[32] The configuration of a data bus to be used will depend on the particular needs. Two key parameters are the number of stations and the total distance transversed (total length of cable required).

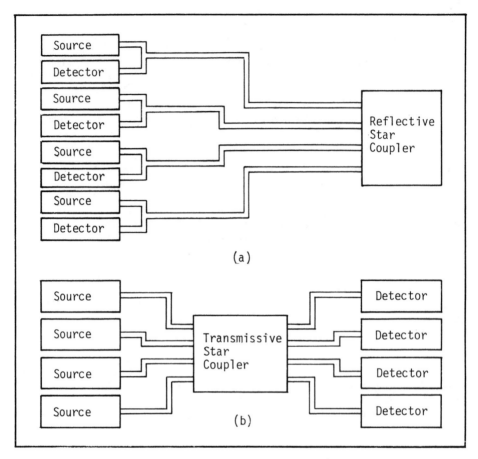

Figure 5.19 - Optical Communications Systems. a) With a Reflective Star Coupler and b) with a Transmissive Star Coupler.

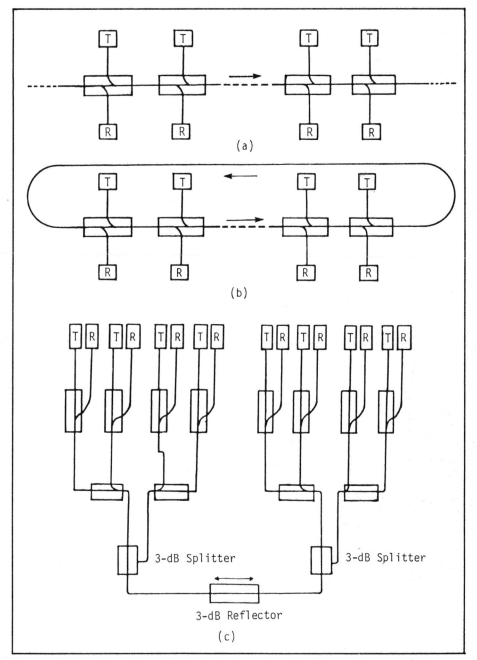

Figure 5.20 - Examples of Optical Data Bus Configurations. a) Line
Bus, b) Ring Bus and c) Tree Bus.

5.7 REFERENCES

1. Stephens, B. and Bunch, T.; *Fiber Optics Use IC Interfaces;* Defense Electronics, pp. 106-107, August 1979.

2. Jones, D.L.; *An Engineering Approach to Fiber Optic Link Design;* Electronic Engineering, p. 65, Mid-April 1980.

3. Botez, D. and Herskowitz, G.J.; *Components for Optical Communications Systems: A Review;* Proceedings of IEEE, Vol. 68, No. 6, pp. 689-731, June 1980.

4. Math, I.; *Basic Optical Data Links;* Electro-Optical Systems Design; September 1977.

5. ITT Electro-Optical Products Division, Technical Note R-2; *Operation of 20 Mb/s Digital Optical Fiber Transmitter/ Receiver Modules;* Roanoke, Va., Jan. 1979.

6. Moulton, P.K.; *Miniaturized, High-Speed, EMI-proof Optical Links Fill the Bill in Computer Systems;* Electronic Design 5, pp. 67-70, March 1, 1980.

7. Stephens, B. and Bunch, T.; *Monolithic Fiber Optic Drivers/ Receivers Use Existing ICs;* Digital Design, pp. 95-96, Oct. 1979.

8. Elmer, B.R., Geddes, J.J. and Biard, J.R.; *Led Driver and Pin Diode Receiver ICs for Digital Fiber Optic Communications;* SPIE Vol. 150, Laser and Fiber Optics Communications, pp. 169-173, Aug. 1978.

9. Hewlett Packard, Application Note 1000; *Digital Data Transmission With Fiber Optic System;* Palo Alto, Ca.

10. Goerne, J. and Keprda, J.; *Optical-Fiber Transmission Systems;* Components Report; Vol. XIII, No. 2, pp. 43-47, 1978.

11. Bash, E.E., Carnes, H.A. and Kearns, R.F.; *Calculate Performance into Fiber-Optic Links;* Electronic Design, pp. 161-166, Aug. 16, 1980.

12. Sorensen, H.O.; *High Speed Low Error Data Transmission with Fiber Optics;* Computer Design, pp. 166-170, March 1979.

13. Wells, P.; *Optical-Fiber Systems for Telecommunications;* GEC Journal of Science and Technology, Vol. 46, No. 2, pp. 51-60, 1980.

14.   Mirtich, V.L.; *Designer's Guide to: Fiber-Optic Data Links-
       Part 1;* EDN, pp. 133-140, June 20, 1980.

15.   Kolodzey, J.S.; *A Noise Analysis Method for Optical Receivers;*
       1977 IEEE International Symposium on Circuits and Systems Pro-
       ceedings, pp. 183-186, Phoenix, Az., April 25-27, 1977.

16.   Miskovic, E.J. and Casper, P.W.; *Noise Phenomena in High-Bit-
       Rate Fiber-Optic Systems;* Electro-Optical Systems Design, pp.
       27-33, May, 1980.

17.   Null, G.M., Uradnisheck, J., and McCartney, R.L.; *Three Tech-
       nologies Forge a Better Fiber-Optic Link;* Electronic Design,
       pp. 65-68, May 24, 1980.

18.   Miller, Eric; *Introduction to Practical Fiber Optics;* IFOC,
       pp. 63-71, September 1980.

19.   Adlerstein, Sid; *Universal Fiber-Optic Receiver-Amp Works Over
       Wide Sensitivity Range;* Electronic Design 3, pp. 23-24, Feb. 1,
       1980.

20.   O'Neil, P.; *Fiber-Optic Receiver Triggers on Edge;* EDN, p. 158,
       June 20, 1979.

21.   Jacobs, I.; *Lightwave Communications-Yesterday, Today, and
       Tomorrow;* Bell Laboratories Record, pp. 2-10, January 1980.

22.   Chown, M.; *Repeaters for Optical Communications Systems;*
       Electrical Communications, pp. 170-179, Vol. 52, No. 3, 1977.

23.   Baldwin, D.L., et al; *Optical Fiber Transmission System Demon-
       stration Over 32 km With Repeaters Data Rate Transparent Up to
       2.3 Mbits/s;* IEEE Transactions on Communications, Vol. COM-26,
       No. 7, pp. 1045-1055, July 1978.

24.   Midwinter, J.E.; *The Development and Application of Optical
       Fibers;* Communications Engineering International, p. 30,
       June 1980.

25.   Jones, J.R.; *Fiber Optics for Telecommunications: Trucking
       and Loop Applications;* IFOC, p. 43, September 1980.

26.   Takasaki, Y.; *Repeater Power Plans for Fiber Optic Communica-
       tion Systems;* IEEE Transactions on Communications, Vol. Com-26,
       No. 1, pp. 195-199, January 1978.

27.   News/International; *53-km Fiber Optic System Needs no Repeaters;*
       Microwave, p.28, September 1978.

28.   Bender, A.; *Fiber-Optic-Bus Status and Applications;* Midcon'
       80, Dallas, Tx, Nov. 4-6, 1980.

29.   Saunfield, J.E. and Biard, J.R.: *A MIL-STD-1553 Fiber Optic
       Data Bus;* Proceedings of the AFSC Multiplex Data Bus Confer-
       ence, Dayton, Oh., 3-5 November 1976.

30.   Coyne, L.J.; *Distributive Fiber Optic Couplers Using Rectangu-
       lar Lighguides as Mixing Elements;* Proceedings FOC' 79, Second
       International Fiber Optics and Communications Exposition, pp.
       160-164, Sept 5-7, 1979.

31.   ITT Electro-Optical Products Division; *Optical Fiber Communica-
       tions;* Technical Note R-8, Roanoke, Aug. 1978.

32.   Reference 11 of Chapter 4.

CHAPTER 6

ANALOG FIBER OPTIC LINKS

Fiber optic transmission systems have been successfully used for transmitting voice and video signals. In certain cases, analog transmission is preferred to digital transmission because it is simpler to implement, does not require coders, decoders, or regenerators, and uses less transmission bandwidth. However, in general, analog transmission requires linear sources with optical power large enough to meet the noise and linearity transmission objectives.

6.1 INTRODUCTION

The use of fiber optics in analog systems is of interest for the transmission of voice and video. Both economy and reliability dictate the analog transmission of television signals and high-fidelity sound over optical fibers with LEDs as transmitters and p-i-n diodes as receivers.[1]

Several analog fiber optic systems have been analyzed in the literature,[2-11] and a variety of them have been successfully used for transmission of voice and video signals. This chapter summarizes the use of fiber optic systems for transmitting analog information. Of course, as fiber optic technology progresses, more analog systems like switched video will become feasible with the development of optical systems operating at longer wavelengths (1.1-1.5 μm).

A goal for future communication systems is the realization of a video phone at the same cost as the voice telephone. Fiber optic transmission is thought to have the potential to attain this communications target. To attain such a system, the following requirements will have to be satisfied;[12]

- Cost of the fiber will have to be at least as cheap as that of the twisted pair cable used for voice signal transmission.

- Video repeater span will have to be as long as that for voice signal transmission (more than ten kilometers).

- Modulators and demodulators for video signals will have to be as inexpensive as those for voice signals.

6.2 ANALOG FIBER OPTIC DATA LINK FUNDAMENTALS

Although most systems applications of fiber optics employ the digital mode, there are several low-cost alternatives using direct baseband modulation for the transmission of analog data. Particularly, analog fiber optic systems can be more effective than their digital counterparts when dealing with slow changing parameters of certain measurement and control systems. Such systems, where V/F and F/V converters are primarily used, are discussed briefly in this section.

6.2.1 Components of a Typical Analog Fiber Optic Link

In the analog fiber optic link shown in Fig. 6.1, the modulator is a voltage-to-frequency converter (VFC) that causes the LED emitter to produce either a high or low optical output level.[13] At the receiving end of the link, the photodiode detects the transmitted light and, in conjunction with additional pre-amplification circuitry, supplies the frequency-to-voltage converter (FVC) input with a stream of on/off TTL-level pulses. The ·instantaneous rate of these pulses is proportional to the instantaneous level of input to the (modulator)

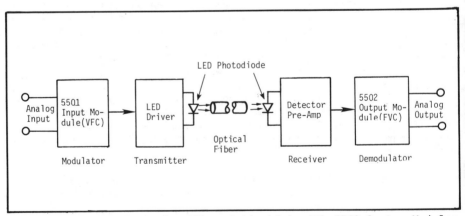

Figure 6.1 - A Basic Analog Fiber Optic Link with 5500 System Modules from Dynamic Measurements.

VFC. In other words, the VFC output drives the optical fiber link with constant amplitude light pulses that occur at rates proportional to the analog input into the system.

6.2.2 Design Considerations [13,14]

The VFC section in Fig. 6.2 is offset at its input to provide a 4.5-MHz carrier frequency for an analog input of zero volts. The off-setting is accomplished by grounding the analog-input-pin and in-jecting approximately 0.25 milliamps of current per MHz of offset desired into the offset-adjust-pin. In the example, 1.1 mA provides the 4.5 MHz offset or carrier. The span can be adjusted over a ±1.5%

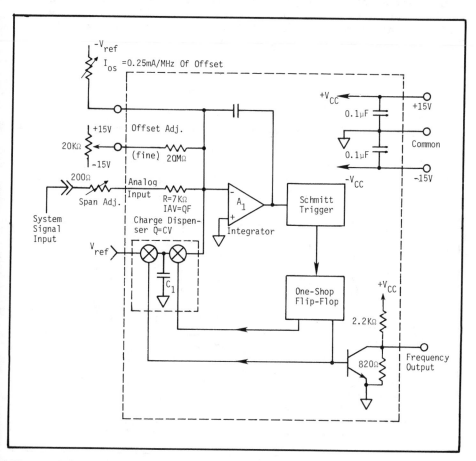

Figure 6.2 - The VFC Section of an Analog Fiber Optic Link.

range by the addition of a 200-Ω pot in series with the analog-input-pin.

The FVC section in Fig. 6.3 is offset to provide an output of zero volts at the carrier frequency. The offsetting is accomplished by applying the carrier frequency at the frequency-input-pin and applying approximately 0.25 milliamps of current per MHz of offset desired into the offset-adjust-pin. In the example, 1.1 mA provides zero volts at the analog-output-pin for a 4.5-MHz input. The FVC span can be adjusted ± 1.5% by connecting a 200-Ω pot from the span-adjust-pin to the output pin.

Another fiber optic link utilizing V/F and F/V converters for DC and low frequency signals is shown in Fig. 6.4. The LM331 produces a pulse stream proportional to the analog input voltage, and modulates the LED coupled to a fiber optic cable. The LH0082 receives the incoming signal, converts it to TTL logic levels, and drives the LM-2907 which performs the frequency to voltage conversion.

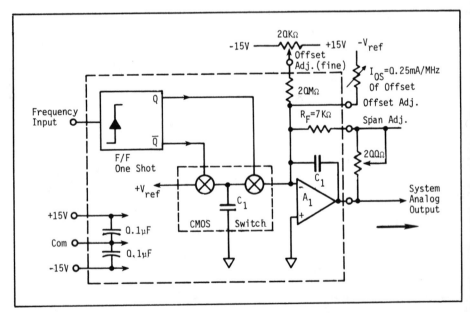

Figure 6.3 - The FVC Section of an Analog Fiber Optic Link.

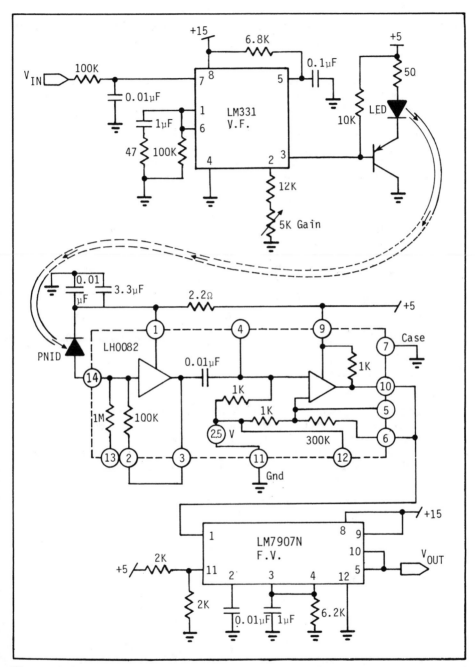

Figure 6.4 - Analog Fiber Optic Data Link Using the LH0082 as an
Interface in the Receiver Section.[14]

6.5

6.2.3 Availability of Packaged Devices and Optical Cables for Analog Transmission [1],[15]

There are at present a variety of packaged devices that can be used to sample analog data and convert it to an FM digital stream. The frequency of the digital data should be at least 10 times the variation rate of the analog signals being sampled. Table 6.1 shows some representative systems that can be assembled for long-distance transmission (to 2 km) and provide the indicated analog bandwidth. Since these systems are based on threshold detection, the DC level is independent of LED power, changes in cable attenuation, or drift in receiver-amplifier characteristics.

Table 6.1 - Bandwidths for V/F - F/V Systems

P/N	V/F-F/V Module Manufacturer	Analog Bandwidth
MDL 24700	Intech	DC to 1 kHz
MDL 24701	Teledyne-Philbrick	DC to 20 kHz
MDL 24702	DMC	DC to 500 kHz
MDL 24703	TRW (A/D-D/A)	DC to 5 MHz

A LED couples twice as much power into a step-index fiber as into a graded-index fiber with the same core diameter and numerical aperture (NA). However, because of their wider modulation bandwidths, graded or semigraded-index fibers are often preferred for wideband analog loops. Table 6.2 summarizes the operating functions of some fiber optic cables suitable for analog applications.

Table 6.2 - Properties and Operating Functions of Fiber Optic Cables in Analog Applications

Manufacturer Fibers	Cables	Core Diameter (μm)	Overall Diameter (μm)	Cladding Diameter (μm)	Single-Fiber Cable Diameter (μm)	Stabilized NA (approx)	Electrical Bandwidth 6 dB (MHz/Km)	Attenuation in cable* (dB/Km)	Typical Coupled Power (μW)
Corning	Siecor	100	400	140	4.8	> 0.3	20	7	150**
Fat Fiber	Belden				3.8				
Corning	Siecor□	100	400	140	4.8	> 0.28	> 100	7	
Wideband Fat Fiber	Belden				3.8				
Quartz Product Corp. QSF-A-200	Belden	200	600	400△	3.8	0.22	25	7	225+
Sumitomo	Sumitomo	80	900	125	3	0.22	130	4	
	ET-8/6004				3.8				
Sumitomo	Sumitomo	80	900	125	3	0.25	90	4	60**
	ET-8/5004				3.8				

*   Selected fiber
** From high-radiance LED, with 50-mA drive
□   Not yet in general production
△   Uses silicone cladding
+   From Spectronics SE3352-004, suitable for large-d fibers up to 300 μm, with 50-mA drive

Source: Adapted from Reference 1.

6.3 LINEARIZATION OF OPTICAL TRANSMISSION DEVICES

When dealing with analog fiber optic systems, improved linearity
of the fiber optic transmission link is required if realistic design
objectives are to be obtained. This is especially true for multichannel signal transmission over systems using repeaters. In such
cases, circuit design techniques for distortion reduction must be
employed.

6.3.1 LED Transmitter Linearity[1]

LED linearity is often critical in analog systems. Unfortunately,
linearity varies from LED-to-LED and also with junction temperature.
High-radiance LEDs offer suitable linearity for analog loops, due to
their low second and third-order intermodulation terms. For example,
in a GTE high-radiance LED, when the envelope of two carriers with
equal amplitudes and frequencies f_1 and f_2, modulate an 80-mA bias by
80%, the second-order term $(f_1 + f_2)$ is 35-dB below one carrier's
amplitude and the third-order term $(2f_1 \pm f_2)$ is 65 dB below. The
third term is low enough to avoid crossmodulation interference in TV
channels with coherent-luminance carriers.

LED linearity can be maintained by simple LED-driver and signal
amplifier circuits like those in Fig. 6.5.[1] The amplifier uses a
complementary pair of transistors, Q_1 and Q_2, to further reduce
second-order distortion. Feedback, via R_1, R_2, and C_1 provides a 75-
Ω input impedance. This amplifier feeds a pair of complementary
drivers (multiple pairs can be used for higher LED currents). The
collector outputs of the drivers are ac-coupled to the LED. The LED
bias can be adjusted by selecting appropriate values of R_3. Diode D_1
protects the LED from inverse transient voltages. Further improvements in transmitter linearity are possible by employing negative
feedback, feedforward, and quasi-feedforward compensation.[3,17]

In a negative feedback arrangement, a photodiode positioned close
to the light source monitors the optical signal and provides the
necessary feedback signal. The amount of distortion compensation is
determined by the feedback loop gain. Although the application of
negative feedback is straightforward, difficulties due to large bandwidth requirements of the feedback circuit may arise at high frequencies.

Distortion compensation using feedforward is achieved through
isolation of the distortion produced in a given nonlinear circuit and
by subsequent injection of the processed error signal back into the
circuit. The advantage of using feedforward is that the bandwidth
of the sampling path is identical to that of the signal channel and

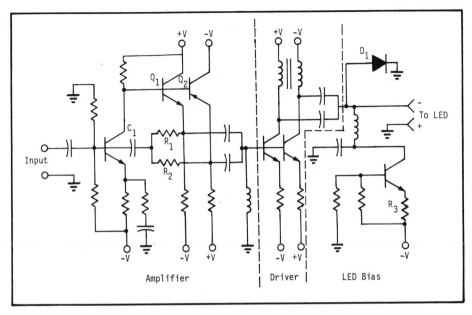

Figure 6.5 - Control of an LED's Linearity by Means of a Three-Stage Transmitter.

the resulting compensation is not dependent on device frequency limi-tations as in the case of feedback.

The optical feedforward system (Fig. 6.6)[3] requires a monitoring photodiode placed adjacent to the LED, an error processing circuit, and a second LED by which a compensating optical signal is generated. This signal is then coupled to the original optical signal through an optical combiner. In this arrangement difficulties are associated with the efficiency of the single fiber optical combiner and, as in the case of conventional feedforward, with amplitude stability of the error signal.

In quasi-feedforward compensation, elements of feedforward and predistortion techniques are combined (Fig. 6.7).[3] The incoming sig-nal S modulates two matched LEDs (LED_1 and LED_2) both of which generate an equal amount of distortion Δ. With the aid of the refer-ence signal path, the distortion from LED_1 is isolated, inverted, and brought to the level required to create a compensating signal equal in amplitude and opposite in sign to the distortion generated by LED_2. Compensation is thus achieved by predistortion of the modulation sig-nal S to S-Δ. By accurate error leveling and proper control of the delays, distortion cancellation can be achieved over a wide range of modulation levels. Quasi-feedforward appears to be the most suitable linearization technique for multichannel transmission.

6.9

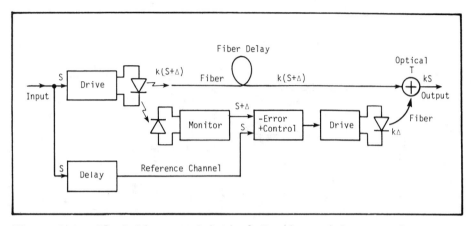

Figure 6.6 - Block Diagram of Optical Feedforward Compensation.

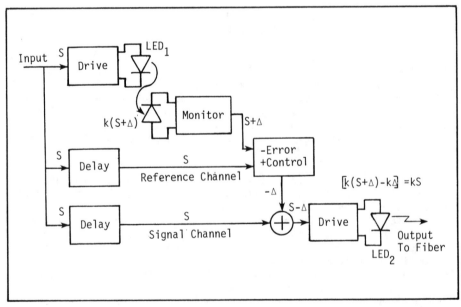

Figure 6.7 - Block Diagram of Optical Quasi-feedforward Compensation.

6.3.2 Laser Transmitter Linearity

Because of its large bandwidth, the injection laser is suitable as a light source for fiber optic CATV application. The simplest and most economical CATV system, however, demands a high degree of linearity on the light output characteristics of the laser.

For analog modulation, a constant current greater than the threshold value is used to bias the laser. The TV carrier signals are superimposed to produce the modulation. The bias current and modulation depth are then adjusted to obtain minimum distortion and maximum signal-to-noise ratio.

A constant laser light output over wide temperature variations can be maintained by utilizing optical feedback. The implementation of this feedback circuit employs conventional, noncritical components and requires only one supply voltage as shown in Fig. 6.8.[19] However, the choice of the photodetector requires careful consideration. The laser's long-term light stability, as controlled by negative feedback, is only as good as the photodetector's long-term light-to-current

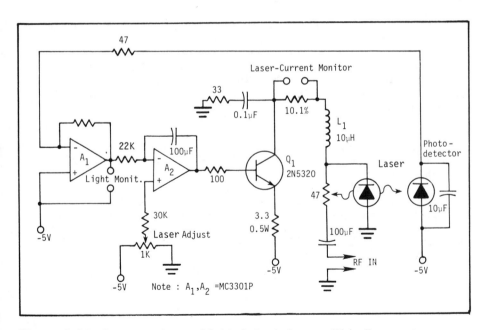

Figure 6.8 - Constant Laser Light Output Over a Wide Temperature Range as Controlled by Negative Feedback.

characteristics. Therefore, a silicon p-i-n diode that exhibits
constant light-to-current conversion over wide temperature ranges
must be chosen.

6.3.3 Microwave Modulation

Microwave modulation can reduce modal noise in analog links.
Experiments with a single-mode laser in an analog system are described in Reference 20, where a 1.6-GHz modulation of 10 milliwatts was
introduced into the bias required to produce one-milliwatt output
from the laser. The laser normally produced most of its output in a
single mode only 0.002-nm wide, but the addition of microwave modulation produced additional modes, each of which was about 0.07-nm wide.
In analog system tests, the microwave modulation improved worst case,
second-harmonic distortion by 6 to 9 dB, and worst case signal-to-
noise ratio by about 3.5 dB. In general, multimode lasers provide
adequate performance in digital systems where linearity is not as
important as in analog operation.

6.3.4 P-I-N Receiver Linearity

If the amplifier and LED-drive circuits are to be substantially
more linear than the LED, the p-i-n photodiode and its transimpedance
amplifier must be more linear than the required overall transmitter
linearity. This approach creates transmitters with adequate linearity
to accommodate three FDM TV channels with coherent luminance carriers.
However, the quasi-feedforward approach is sensitive to gain and to
changes in the LED differential linearity, so is also temperature
sensitive.

In a simplified p-i-n photodiode receiver as shown in Fig. 6.9,[1]
a FET (Q_1) is a source follower, with the photodiode connected between
gate and source. The photodiode uses a 90-Volt bias for low capacitance. The result is a 12-dB per octave response that the equaliza-
tion circuits, Q_5 and Q_6, correct. The source follower drives the
cascode stage, made of FET Q_3 and UHF transistor Q_4. The low-impedance input of Q_4 minimizes the Miller effect in amplifier Q_3. This
receiver provides 111-dB S/N power density at 20 MHz, for incident
light power of 3 to 4 µW.

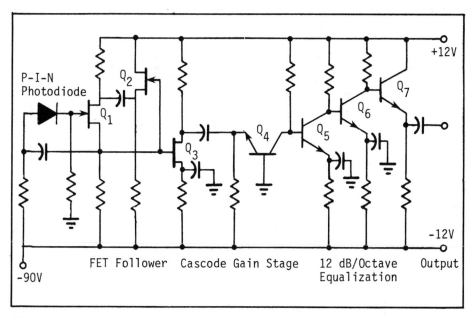

Figure 6.9 - A P-i-n Photodiode-based Receiver Using FET Follower and Bootstrapping Technique to Reduce the Effects of Gate-to-Sink Capacitance.

6.4 FIBER OPTIC ANALOG VIDEO TRANSMISSION

Just as in telecommunications, the advantages of fiber optics in cable television networks are low attenuation, wide bandwidth, immunity to electromagnetic interference and lightning, light weight, small cross-section, and ease of installation. However, a closer examination shows that universal use of fiber optics as a replacement for coaxial cable in conventional CATV network is limited by state-of-the-art optoelectronic components.[11] The most stringent requirements in fiber optic analog links are posed by high-quality analog video transmission.

6.4.1 Commercial and Studio-Quality Video

In terms of the transmission objectives used for broadcast-quality television signals, one should differentiate between commercial and studio-video qualities. The relevant transmission objectives to consider in the design of fiber optic transmission systems are: (1) weighted signal-to-noise ratio (SNR), (2) differential gain (DG), and (3) differential phase (DP). Transmission of commercial-quality video requires SNR=53 dB, DC=1.4 dB , and DP=5 degrees. Transmission of studio-quality video is more difficult to achieve because it requires SNR=70 dB, DG=0.01 dB, and DP=0.5 degree.[10]

6.4.2 Single-Channel Video and Audio Transmission

The analog transmission of single-channel, high-quality, color video and audio can be accomplished using either coaxial cable or fiber optics. Applications for a point-to-point single channel of video and audio are found in broadcasting (studio or remote cameras), teleconferencing, surveillance monitoring and satellite entrance links.[21] Typical input/output is a 1-Volt p-p signal through 75 Ω (baseband signal). This signal may be transmitted baseband, or frequency modulated, or higher FM carrier. FM transmission requires additional conversion from and to baseband.

A fiber optic baseband video transmission system is functionally similar to other cable systems. It consists of a line driver (transmitter), cable, and line receiver. However, the way in which these components are implemented in fiber optic systems differs drastically from conventional cable systems. A block diagram of a fiber optic transmission system is shown in Fig. 6.10.[22] The transmission medium is completely dielectric, and therefore does not conduct electric currents. At the transmitter end, a composite video signal from a camera or VTR is fed through a 75-Ω coaxial patch to a voltage-to-

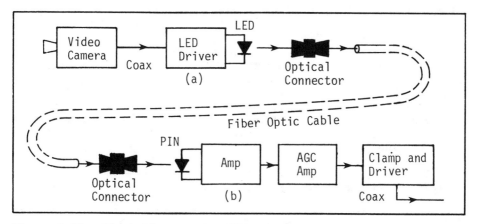

Figure 6.10 - Block Diagram of a Fiber Optic Video Transmission System. a) Transmitter end. b) Receiver end.

current converter to drive the light source. The fiber optic system, due to the processing performed at the receiver, appears transparent to the electrical signal, and in a properly designed system, results in minimal degradation of the video quality.

6.4.3 Multi-Channel Video Channel Transmission

Although the transmission of a single video channel over a few Km is relatively straightforward, the universal use of fiber optics as a replacement for coaxial cable in a conventional multichannel FDM network (e.g. CATV) is still limited by the state-of-the-art optoelectronic components, especially by the performance limitations of readily available light sources. However, in the future, continued improvements in the performance of optoelectronic transmitters and auxiliary fiber optic hardware will ensure cost effectiveness in CATV applications.[11,23]

At the present time, there are three major approaches that can be applied to video trunking: (1) intensity modulation using analog baseband amplitude information, (2) intensity modulation using frequency modulated subcarriers, and (3) binary PCM digitized video. In short distance trunks, economics clearly favors analog intensity modulation.

6.4.4 A 6-Km Installed Video Trunk System

A fiber optic trunk system that has been installed in Joplin, Missouri is described in References 3 and 11. The system carries three video signals from the Local Origination Studio (LOS) to the Headend (HE), and a return video signal from the HE to the LOS. The total distance is 6 Km. This return video channel is used to bring programming from a satellite earth station to the LOS. The system block diagram is shown in Fig. 6.11.[11] Each fiber of the four-fiber cable carries a single video channel.

The audio signal modulates the audio subcarrier and the combined video/audio signal frequency modulates the optical source, a light

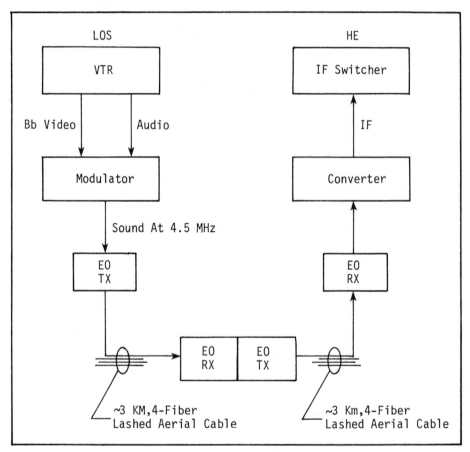

Figure 6.11 - System Block Diagram Showing One of the Three Identical Channels from the LOS to the HE. The Fourth, i.e., Return Channel is Similar.

emitting diode. The bandwidth of the combined signal is 5 MHz and
the frequency deviation of the FM carrier is ± 12 MHz. At the HE,
the silicon avalanche photodiode (APD) converts the intensity-modulat-
ed optical signal into an FM electrical signal which is demodulated to
provide the combined video/audio signal. This is fed to an IF modulator and then to an IF switcher which routes the signal to the appropriate channel converter. The repeater is a combination of an electro-
optic (EO) receiver and transmitter. The modulation technique used
(FM/IM) is distinguished by its high linearity and temperature stability. This guarantees a high signal-to-noise ratio and is insensitive
to variations in the attenuation of the transmission route.

The block diagram of the optoelectronic transmitter of the 6-Km
video system is shown in Fig. 6.12a.[11] A 20-MHz carrier is frequency
modulated by the 5-MHz bandwidth input signal. The frequency modulation is performed by a push-pull modulator. A differential amplifier
supplies signals with opposite phase which frequency modulate oscilla-
tors with center frequencies at 130 MHz and 150 MHz. A mixer combines
the two modulated signals into one resulting carrier which then modulates the intensity of the light source, a doubleheterostructure high-
radiance Burrus LED.

Figure 6.12b[11] is a block diagram of the optoelectronic receiver.
The modulated light signal is converted by the photodetector into an
electric current. This signal is amplified, limited and demodulated.
The pre-emphasis network in the transmitter and the de-emphasis network in the receiver have been designed to optimize the video S/N.
For the detector, a silicon avalanche photodiode is used which results
in some reduction in the required received optical power.

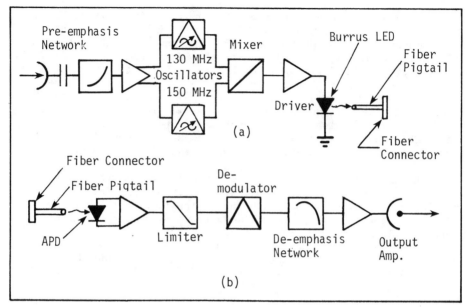

Figure 6.12 - Block Diagram of Optoelectronic Transmitter a) and
Receiver b) of Fiber Optic Trunk System for Analog Video Transmission.

6.5 REFERENCES

1.   Tweedy, E.P.; *Put Analog Transmissions on a Fiber-Optic Loop;* Electronic Design, pp. 171-176, August 16, 1980.

2.   Albanese, A.; *Applications of Lightguide Systems to Analog Transmission;* 1980 Wescon Professional Program, Anaheim, Ca. September 16-18, 1980.

3.   Straus, J. and Szentesi, O.I.; *Linearization of Optical Transmitters by a Quasified-Forward Compensation Technique;* Electronics Letters, Vol. 13, No. 6, 7 March 1977.

4.   Marvin, L.A. and Kao, C.K.; *Fiber Optics Links: CATV System Design for the Future;* TV Communications, Vol. 12, pp. 20-25, December 1975.

5.   Szentesi, O.I., and Szanto, A.J.; *Fiber Optics Video Transmission;* SPIE, Vol. 17, pp. 151-156, 1976.

6.   Horton, R.; *Analog F.D.M. Telephony Transmission in Optical Fibers: The Limitations of Intermodulation Distortion;* Electronics Letters, Vol. 12, No. 15, pp. 386-387, July 1976.

7.   Holden, W.S.; *An Optical-Frequency Pulse-Position-Modulation Experiment;* Bell System Technical Journal, Vol. 54, No. 2, pp. 285-296, February 1975.

8.   Jones, J.R.; *An Analog Optical Communications Channel Using Wideband Subcarrier FM;* National Telecommunication Conferences, p. 37. 4-1, 1976.

9.   Straus, J.; *Analog Signaling in Optical Fiber Systems;* Proceedings of the National Electronics Conference, Vol. 23, pp. 49-53, Chicago, October 1979.

10.   Albanese, A., and Lenzing, H.E.; *IF Lightwave Entrance Links for Satellite Earth Stations;* International Conference on Communications IEEE, Boston, June 1979.

11.   Szentesi, O.I., Knegler, E., and Petty, W.D.; *A Fiber Optic Trunk System for Analog Video Transmission;* INTELCOM' 79, Dallas, Texas, Feb. 26 - March 2, 1979.

12.   Takasaki, Y., Nakagawa, J., and Koya, M.; *New Fiber Optic Analog Baseband Transmission Plan for Color TV Signals;* IEEE Transactions on Communications, Vol. COM-26, No. 6, pp. 902-907, June 1978.

13. Dynamic Measurements Corp., Application Techniques AT-802; *Attaining High Accuracy in Fiber Optic Analog Data Links;* Winchester, Ma.

14. Miller, Eric; *A Versatile Hybrid Fiber Optic Receiver;* Electronic Design Conference for Telecommunications Systems Designers, Telcom Design 1980, Woburn, Ma, Sept. 30 - Oct. 2, 1980.

15. Medved, D.B., and Keating, J.; *On the Transmission of Linear High Speed Analog Data Over Fiber Optics;* FOC' 80, San Francisco, September 16-18, 1980.

16. Straus, J., and Szentesi, O.I.; *Linearized Transmitters for Optical Communications;* Proceedings of IEEE Conference on Circuits and Systems, p. 288-292, Phoenix, Az., April 1977.

17. Freeborn, J.C.; *Speed Up Digital Data Transmission Over Fiber-Optic Links with Analog Feedback;* Electronic Design 3, pp. 74-76, Feb. 1, 1979.

18. Chan, D, and Yuen, T.M.; *System Analysis and Design of a Fiber Optic VSB-FDM System for Video Trunking;* IEEE Transactions on Communications, pp. 680-686, July 1977.

19. Svacek, J.F.; *Transmit Feedback Techniques Stabilize Laser-Diode Output;* EDN, pp. 107-111, March 5, 1980.

20. Miskovic, J.; *Effects of Microwave Modulation of ILD Spectral and Performance Characteristics;* FOC' 80, San Francisco, Sept. 16-18, 1980.

21. Chen, F.S., Karr, M.A., and Shumate, P.W.; *Laser Transmitters for 70-MHz Entrance Link;* The Bell System Technical Journal, Vol. 58, No. 7, pp. 1617-1629, Sept. 1979.

22. Cerny, R.A., and Witkowicz, T.; *Economic Considerations for Single Channel Video and Audio Transmission Over Fiber Optic and Coaxial Cable;* Proceedings of FOC' 79, pp. 149-151, Chicago, Sept. 5-7, 1979.

23. Dabby, F.W., and Chesler, R.B.; *Multiple Video Channel Transmission on a Single Optical Fiber;* Proceedings, SPIE, Washington, DC, March 28-29, 1978.

CHAPTER 7

SELECTION CRITERIA FOR FIBER
OPTIC LINKS AND ASSOCIATED COMPONENTS

As fiber optic and other optical components increase in number and improve in quality, it is important for both designers and users to apply proper criteria, when selecting and fitting these components or systems to an application, in order to obtain higher performance and insure economical advantages.

7.1 INTRODUCTION

Considerable effort has gone into the design of various fiber optic digital and analog systems. Therefore, the technology as it applies to that field is well understood. Also, many cost justifications have been made for the use of fiber optics in digital voice and data systems. The economic and performance advantages of digital video trunking for CATV have also been shown for fiber optics over very long distances. However, for short-to-moderate distances digital techniques become prohibitively expensive.

When considering economical advantages in selecting a transmission technology, it is necessary to define those system elements which are relevant to one-for-one system replacement. The designer must also be aware of the additional benefits provided by the competing technology.[1]

7.2 OVERALL SYSTEM CONSIDERATIONS

At the outset, the various pieces that make up a fiber optic system can be optimized independently. The effort is directed toward the selection of the lowest-loss fibers, the most powerful transmitters, and the most sensitive receivers. However, there are a few areas where interactions lead to compromise.

7.2.1 System Specifications

In determining the optical specifications for any system, the two most important parameters are the optical loss the system will tolerate and the minimum bandwidth required for the system. To determine the maximum optical loss, the losses of all components in the system must be determined. The loss that is important to determine is the maximum loss of the fiber cable. To determine this, all other losses are subtracted from the system gain. The known losses are the source-to-fiber coupling loss, transmit connector loss, receive connector loss, fiber-to-photodetector coupling loss, and system-loss margin. System-loss margin consists of manufacturing tolerances, temperature effects and aging of the optical components.[2]

Assigning some hypothetical values to the system gain and to the main loss mechanisms, the system length may be calculated as shown below:

- Total System Gain: _____ 58 dB
- Source-to-Fiber Coupling Loss: _____ 3.5 dB
- Transmit Connector Loss: _____ 2.5 dB
- Receive Connector Loss: _____ 2.5 dB
- Fiber-to-Photodetector Coupling Loss· _____ 2.5 dB
- System-Loss Margin: _____ 7 dB
- Fiber-Cable Loss (including splices): _____ 40.0 dB

Assuming a low-loss fiber (2.5 dB/Km), system lengths can be up to 16 Km for the above assigned values of losses.

7.2.2 Fiber Optic Baseband vs. Coax Baseband Systems

The most straightforward method of transmitting composite video over coaxial cable is the direct interconnection of equipment, i.e., video camera-to-monitor. Twin-lead coax (video pair) is normally used to eliminate *hum* problems. In cases where external noise problems are not severe, this method is adequate up to distances of a few hundred meters. Beyond that distance, expensive equalizing equipment is required. The greater the distance, the more expensive the equipment becomes. In many cases, signal degradation becomes sufficiently severe after distances of over a kilometer such that FM transmission over coaxial cable may be required.

It is to be noted that a fiber optic baseband system requires an initial electro-optic conversion from 75 Ω and back. With today's technology such a system, purchased off-the-shelf, can deliver a studio-quality signal over distances of up to 3 Km, with no intermediate repeaters. The upper frequency of the baseband coax described herein is limited to 8 MHz. The corresponding baseband fiber optic system is capable of yet higher resolution – over 10 MHz.[1,3] In order to send both baseband video and audio over a video pair coax, a separate audio modulator/demodulator function is required, plus the attendant power supplies.[1,4]

7.2.3 Fiber Optic Baseband vs. Coax FM Systems

An alternate method for metallic cable video trunking is frequency modulation (FM) which offers lower noise, but is higher in cost for single-channel transmission. Typical frequency response of the coaxial FM video system is 30 Hz to 4.2 MHz.[1,3] Table 7.1 shows cost comparisons for coaxial baseband and coaxial FM systems versus fiber optic baseband systems.[3] It is apparent that baseband transmission of video and audio over fiber optics has a distinct cost advantage for distances in excess of one kilometer. In shorter distance applications, the lower cost of a coaxial cable system may well be offset by the inherent transmission problems of metal cable: ground loops, noise (EMI, hum) and susceptibility to lightning damage. Fiber optic baseband video transmission also appears economical when compared to multichannel FM coaxial cable systems.

Table 7.1 - Cost Comparisons for Coaxial Baseband
and Coaxial FM Systems Versus Fiber-Optic Baseband System[3]

	HARDWARE	≤ 1 km	≤ 2 km	≤ 3 km
COAXIAL (BASEBAND)	CABLE (VIDEO PAIR) @ $1.25/m	$ 1250	$ 2500	$ 3750
	AUDIO MODULATOR	395	395	395
	POWER SUPPLY (A.M)	200	200	200
	AUDIO DEMODULATOR	395	395	395
	POWER SUPPLY (A.D)	200	200	200
	LINE EQUIPMENT			
	(DRIVER, RECEIVER, EQUALIZER)	1600	4475	8500
	TOTAL	$ 4040	$ 8165	$ 13440
COAXIAL (FM)	CABLE @ $1.30/m	$ 1300	$ 2600	$ 3900
	VIDEO/AUDIO FM MODULATOR	1550	1550	1550
	VIDEO/AUDIO FM DEMODULATOR	1850	1850	1850
	LINE AMPLIFIER (1300m SPACING)		1200	2400
	LINE AMPLIFIER POWER SUPPLY		600	600
	TOTAL	$ 4700	$ 7800	$ 10300
FIBER-OPTIC	CABLE	$ 1700	$ 3800	$ 6450
	TERMINAL EQUIPMENT	4000	4000	4000
	TOTAL	$5700	$ 7800	$ 10450

7.3 FIBER OPTIC COMMUNICATION SYSTEM COMPONENTS

The essential elements of any light transmission communication system are the transmitter and receiver connected by optical fiber cabling and fiber optic connectors. For these components, it is important to understand and specify their functional features in such a way as to insure balanced interaction and optimum link operation.

7.3.1 Receiver and Transmitter Interaction[5]

An important interaction between the receiver and the transmitter is the transmitted pulse duty cycle. Optical transmitters tend to be peak-power limited, and thus the amount of energy which can be packed into an optical pulse increases linearly with the width of the pulse. Also, high speed transmitters are easier to build with full-duty cycle pulses. For a given energy in a received pulse, the receiver performance is best for narrow received pulses. By analyzing the trade-off between energy per pulse and pulse width, it can be concluded that for fibers with modest amounts of dispersion, the optimal transmitted pulsewidth is full-duty cycle or close to full-duty cycle. Thus, high-speed system transmitters are designed for a nonreturn-to-zero format.

From the theoreticians point of view, the purpose of a receiver in a communications system is to process information from a carrier with a minimum requirement on the carrier level. The practical designer may have another view of the job because of real world constraints.

The most significant parameter which causes compromises in the design of practical receivers is dynamic range. In the laboratory, the optical input power to the detector can be adjusted to the low levels which are appropriate for testing the sensitivity of the receiver. In the real system, the receiver must not only accommodate the minimum detectable signals, but also signals which are significantly larger. This is caused by variations in repeater spacings, variations in fiber losses, connector and splice losses, transmitter output, etc.

7.3.2 Fiber Optic Cables for Practical Applications

The vast majority of cables which are being manufactured for telecommunication applications contain fibers of the multimode, graded-index type. Single-mode, step-index fibers are still in their infancy and are being used mainly in research and development.

Multimode, step-index fibers are mainly used in non-telephone systems because of lower bandwidth requirements.

Since both attenuation and bandwidth vary significantly with wavelength, when they are specified or measured, the wavelength(s) or wavelength range required must be considered.[6] Most multimode graded-index fibers are optimized for transmission at a particular wave-length, typically 0.8-0.9 μm. Fibers having attenuations of 2-5 dB/Km and BW products of 200-1000 MHz·Km are readily available. Recently, fibers have become available which are optimized for a wavelength and are referred to as having *Extended Sprectral Response*. They allow operation at any wavelength in the range, typically 0.8-1.3 μm, with attenuation of less than 3 dB and BW products greater than 400 MHz·Km.[7]

7.3.3 Environmental Criteria for Fiber Optic Cables

Among the most important environmental considerations in fiber optic cable selection are the following:[8]

- The range of temperatures which the cable will experience over its lifetime.

- Potential mechanical abuse, i.e., impact, crushing, flexing, tension caused by installation, wind, and ice loading.

- Potential ingress of water into the cable and the resultant weakening of fiber strength due to stress corrosion or fiber damage due to ice formation.

- Potential damage due to gnawing rodents.

As in conventional wire cabling, the principal concern is environmental extremes rather than normal conditions. For example, it is most important to know the minimum and maximum allowable tempera-tures. For most of the environmental factors listed above, the mate-rials and designs which have been successfully employed for a great number of years in the manufacture of metallic cables are applicable to optical cable without major modifications. However, standardized specifications for optical cables are required that will assist both users and manufacturers.

7.3.4 Fiber Optic Cable Standardization

The ease of choosing the best fiber optic cable for an applica-tion by obtaining meaningful comparisons of specifications depends on

7.6

the cable manufacturers' abilities or inclination to describe their products with some sort of standard terminology.[9] Thus, unstandardized specifications can complicate comparisons of important cable parameters.[10]

Fortunately, attempts to standardize are being made both on an international and national level[7] as shown in Table 7.2. The most prominent international organizations working in this field are CCITT (Study Group XV) and IEC (Technical Committee No. 46E). On the national level, the most active groups have been the military, the NBS and the EIA (P6). In addition to participating in the above mentioned activities, some of the more influential independent organizations (e.g. AT & T, REA) have promoted standarization. It is expected that other groups will be involved in such activities in the future.

Table 7.2 - Some Organizations Working on Fiber Optic Standards

International
International Telegraph and Telephone Consultative Committee (CCITT),   Study Group XV – Working Party on Optical Fibers.
International Electrotechnical Commission (IEC): Technical Committee No. 46; Cables, Wires and Waveguides for Telecommunication Equipment (Sub–Committee 46E–Fiber Optics)
National
Electronic Industries Association: Parts Division, Committee 6 on Fiber Optics (Working Group P6.7 on Cables).
Government–Department of Defense (DOD), MIL and FED Standards   –Department of Commerce/NBS

In order to determine standards of performance for fiber optics, test procedures have to be established simulating the environmental conditions to be encountered by the particular cable type. Table 7.3 lists a number of optical and mechanical cable performance parameters which have to be determined by standard test methods.

7.3.5 Connector Specifications

Connector designs using mechanical means to achieve accurate fiber alignment are provided with precision sleeves, V-shaped grooves, and overlapping rods. Special fixtures are required to locate fiber ends and in certain cases grinding and polishing techniques are applied.

Table 7.3 - Optical and Mechanical Performance Tests
for Fiber Optic Telecommunication Cables

1. Optical Performance		
Attenuation		DOD-STD-1678, M-6020
Bandwidth, Pulse Spreading		, M-6050
Numerical Aperture		, M-8040
Refractive Index Profile		, M-6090
Far End Crosstalk		, M-6060
2. Mechanical Performance		
Physical Dimensions	• Fiber	, M-1010
	• Cable	
Tensile Strength	• Structural	, M-3010
	• Short Term	
	• Long Term	
Bending	• Short Term	
	• Long Term	
Twist Bend		, M-2060
Cold Bend		, M-4010
Flexing		, M-2010
Temperature		
(shipping, installation, operating)		
	• High Temperature	, M-4010
	• Low Temperature	, M-4010
	• Cycling	, M-4020
Crush		, M-2040
Ice Crush		, M-4050
Impact		, M-2030
Humidity		, M-4030
Wicking		, M-8020
Flammability		, M-5010

In the case of detachable fiber optic connectors, it is necessary to deal with losses in the system, not just at the connection interface. For example, if the connector couples power into modes with high attenuation, a significant loss - extrinsic to the connector interface - is induced in a long receiving fiber. Table 7.4[11] shows a set of specifications that must be given proper attention when selecting fiber optic connectors.

Table 7.4 - Connector Specifications

A detachable connector must:

- Be field-terminable within a few minutes.

- Require no chemical processing, epoxy, grinding, polishing, or special fixtures.

- Accommodate waveguides, the physical and optical properties of which may vary somewhat.

- Maintain the integrity of fiber ends in the connect-disconnect process.

- Have less than a 1.0 dB repeatable coupling loss.

- Be of such character that minute-sized particles are not critical.

- Be constructed in such a way that collective tolerances of assembled connector components are not predicated on critical displacement or angular alignment of the fibers.

- Meet environmental requirements.

- Be capable of production at a reasonable cost in high volume.

7.4 OPTOELECTRONIC MODULES AND DEVICES SELECTION

Specifications on the data sheets are typical of an average device. Each device must be considered individually and driven accordingly. Selection tips for optoelectronic devices such as sources, photodetectors and optical isolators have already been given in chapters 2 and 3. In this section, some additional information is provided regarding the selection of practical driving circuits and photodetectors.

7.4.1 CW Laser and LED Driver Circuits

Table 7.5 is a summary of characteristics and tradeoffs for various driving circuit applications using three basic configurations: (1) single ended digital, (2) linear analog and (3) differential digital.[12] It is suggested, however that they be considered as starting points rather than as rigid rules. The final circuit for any application should be achieved through standard engineering evaluation and practice.

Table 7.5 - Summary of Characteristics and Tradeoffs
for Various CW Laser and LED Driver Circuit Configurations

APPLICATION	CIRCUIT CONFIGURATION		
	SINGLE-END DIGITAL	LINEAR ANALOG	DIFFERENTIAL DIGITAL
High Precision Current Control	Poor	Excellent	Excellent
To Drive Laser	Poor	Excellent	Excellent
To Drive LED	Excellent	Excellent	Good
Low Cost	Excellent	Fair	Fair
High Efficiency	Good	Fair	Fair
Low-rate Digital	Excellent	Poor	Excellent
High-rate Digital	Fair	Poor	Excellent
T1 Telephony	Excellent	Poor	Excellent
T3 Telephony	Fair	Poor	Excellent
Baseband Video	No	Excellent	No
TDM PCM	Excellent	Poor	Excellent

7.4.2 Requirements for Practical Fiber Optic Detectors[13]

The performance and cost constraints upon fiber optic communications systems are well matched by the various detectors for such systems. Generally, silicon p-i-n diodes work best for short-haul, low-cost systems, and silicon avalanche photodiodes (APD) are preferred when systems have bandwidth-distance products of up to 1000 MHz·Km, and when the detector's cost is not a prime consideration. APDs are most useful when the signal is not the predominant noise source. For systems requiring bandwidth-distance products of approximately 5,000 MHz·Km, quaternary III-V APDs coupled to GaAs FETs will achieve a wavelength of 1.3 μm. Photodiodes of silicon, both the p-i-n and APD types, are about the only detectors that presently satisfy all requirements of fiber-optic communications systems.

Among the main requirements that a practical fiber optic detector must meet are:

- High responsivity in the emission range of the available optic source.

- Fast response.

- Low noise.

- Insensitivity to temperature.

- Practical coupling to optical fiber.

- Long operating life.

- Reasonable cost.

Table 7.6 lists the important parameters of commercially available silicon p-i-n photodiodes for the 0.8 to 1.06 μm range. Table 7.7 lists silicon APDs and Table 7.8 lists commercially available, long wavelength detectors.[13]

Table 7.6 - Commercially Available P-i-n Photodiodes

DEVICE	Respon- sivity (A/W) @ λ_o (nm)	3dB band- width (MHz)	Dark Current (nA @ V)	Active area (mm)^2	PACKAGE
RCA C30807	0.6 @ 900 0.15 @ 1060	60	2 @ 10 10 @ 45	0.8	TO-18
RCA C30808	0.6 @ 900 0.15 @ 1060	45	5 @ 10 30 @ 45	5.0	TO-5
RCA C30809	0.6 @ 900 0.15 @ 1060	25	25 @ 10 70 @ 45	50.0	TO-8
RCA C30810	0.6 @ 900 0.15 @ 1060	20	80 @ 10 300 @ 45	100.0	RCA 25-mm
RCA C30822	0.6 @ 900 0.15 @ 1060	40	10 @ 10 50 @ 45	20.0	TO-8
RCA C30831	0.6 @ 900 0.15 @ 1060	60	1 @ 10 10 @ 45	0.2	TO-18
RCA C30812	0.6 @ 900 0.15 @ 1060	25	20 @ 20 30 @ 200	5.0	TO-5
UDT PIN-020A	0.42 @ 850	75	0.5 @ 10	0.2	TO-18
EG&G SGD-040	0.5 @ 900	120	5 @ 100	12.6	TO-46
Thompson CSF TCO-202	0.4 @ 830		2 @ 10	0.5	Special/ F/O
HP5082-4205	0.5 @ 770	360	0.15 @ 10	0.02	Pill
BNR D-5-2	0.55 @ 840	120	1 @ 45	0.012	TO-1811/ F/O
RCA C30900E	0.6 @ 900	60	10 @ 10 100 @ 90	5.0	TO-5
NEC LSD 39A	0.3 @ 633	1800	0.2		
NEC LSD 39B	0.3 @ 633	900	0.1		

Table 7.7 - Commercially Available Silicon APDs

Device	Responsivity (A/W) at λ_o (nm)	3dB Band-width (MHz)	Bias (V)
RCA C30902E	77 @ (830), 65 @ 900	720	180-250
RCA C30884	63 @ (900), 8 @ 1060	400	230-330
RCA C30817	75 @ (900), 18 @ 1060	200	275-425
RCA C30895	63 @ (900), 20 @ 1060	200	300-425
RCA C30872	37 @ (900), 29 @ 1060	200	275-425
RCA C30916E	50 @ (900), 12 @ 1060	200	275-425
TI TIXL-56	25 @ (900)	10	160
TI TIXL-451	20 @ (900)	200	100

Table 7.8 - Commercially Available Long-Wavelength Detectors

Device	Responsivity (A/W) @ λ (nm)	3dB B.W (MHz)	Dark Current (nA @ V)	Active Area (mm)^2
Optitron GA-1 (Ge APD)	11 @ 1300 @ 44.7V	450	10.000@ 44.7	0.1
Mitsubishi PD-7000 (InGaAsP P-i-n)	0.6 @ 1300	600	100 @ -15	8×10^{-3}
Judson Infrared J-16 (Ge p-i-n)	0.4 @ 1500	200		

7.5 PACKAGING ASSEMBLIES AND COMPATIBILITIES

Low-cost, all-plastic fiber optic assemblies have already appeared that are specially designed for high-volume usage, particularly over short distances. Although their specifications are certainly not extraordinary, these assemblies do provide some features that increase system design flexibility.

7.5.1 All-Plastic Fiber Optic Source and Detector Assemblies

The new assemblies that are available employ plastic fiber having typically a 0.37-mm core diameter and a large numerical aperture of 0.53. Consequently, coupling efficiency is high, approximately 20 to 25%. Due to large cost diameter systems, costs are relatively low, since precision alignment is not required. This means that low-cost plastic fiber optic connectors and economical large area planar surface-emitting diodes may be used. In addition, plastic fiber makes termination easier, as it is very ductile and highly durable. Usually, the plastic source and detector assemblies incorporate an integral fiber optic cable pigtail.[14] One end of this pigtail is permanently aligned to the emitter or detector chip for optimum optical coupling efficiency. The other end is terminated with a plastic fiber optic connector or ferrule. The user, then, need only be concerned with a simple fiber-to-fiber connection.

7.5.2 Special DIP for Fiber Optics

In a plastic DIP, lead frames carrying the mounted electro-optical device are assembled in a plastic molded case, which is then epoxy-filled to produce a solid assembly (as shown in Fig. 7.1). Standard high-volume lead frame assembly techniques are used for chip mounting and wire bonding.[14] In addition, the DIP's internal design enables simple precise self-alignment of the fiber optic relative to the emitter or detector chip. The back three leads of the package make electrical connection to the electro-optical device, while the front three leads provide additional mechanical support. For improved heat sinking, an emitter chip makes thermal contact with two of the leads.

7.5.3 TTL-Compatible Packaging

A number of manufacturers offer fiber optic kits to help simplify experimentation and assembly. System designers can construct

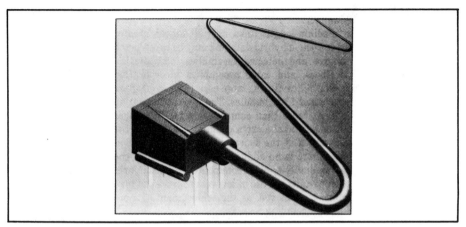

Figure 7.1 - Special DIP for Fiber Optics.

systems which are both TTL-and MOS-compatible. Of course, a kit cannot produce a *saleable* system, but it can bring an experimental fiber
optic link to about 85% of the way toward a final commercial design.
Still needed are environmental protection details such as temperature
compensation and the obvious repackaging for production economy.[15]

    The best examples of the movement away from designer kits toward
preassembled and pretested fiber-optic transmitters and receivers are
the new data links intended as plug-in units for computer OEMs.
These complete systems are totally compatible with TTL circuitry.
Some, like the high and low speed data links from 3M, mate readily
with standard PC-board headers. Others, like the transmitter and
receiver modules from Honeywell's Spectronics Division, can be solder-
ed onto PC boards.[16]

7.6 REFERENCES

1. Cerny, R., and Witkowicz, T.; *Fiber-Optic Baseband Video Systems Can Provide Economic Advantages Now;* Communication News, Sept. 1979.

2. Bunker, N.S., Cheung, H.W., and Ester, G.W.; *Design Considerations for a 44.7 Mb/s Fiber Optic Transmission System;* 1980 Midcon Professional Program, Dallas Covention Center, November 4-6, 1980.

3. Reference 22 of Chapter 6.

4. Dineson, M.A.; *Broadband Coaxial Local Area Networks Part 1: Concepts and Comparisons;* Computer Design, p.12, June 1980.

5. Personick, D.; *Receiver Design for Optical Fiber Systems;* Proceedings of the IEEE, Vol. 65, No. 12, pp. 1670-1678, Dec. 1977.

6. Bark, P.R., and Liertz, H.M.; *Design of Optical Waveguides and Cables for Practical Applications;* 1980 Midcon Professional Program, Dallas Convention Center, Nov. 4-6, 1980.

7. Bark, P.R. and Lawrence, D.O.; *Emerging Standards in Fiber Optic Telecommunications Cable;* Proceedings of the SPIE's Technical Symposium East 1980.

8. Hudson, M.C. and Dobson, P.J.; *Fiberoptic Cable Technology;* Microwave Jorunal, pp. 46-53, July 1979.

9. Ranada, D.; *Fiber-Optic-Cable Specs Require Careful Scrutiny;* EDN, pp. 101-104, March 20, 1979.

10. Georgopoulos, C.J. and Athanasiadis, N.; *Fiber Optics vs. Conventional Transmission Lines in Distance Communication Systems;* Second International Conference on Information and Systems, Vol. I, pp. 305-312, University of Patras, Greece, July 9-13, 1979.

11. Duyan, Jr., P.; *Fiber Optic Interconnection;* Industrial Research/Development, pp.68-71, December 1979.

12. Adair, R.; *CW Lasers and LED's;* Application Note A/N 101, Laser Diode Laboratories, Inc., N.J.

13. Reference 27 of Chapter 2.

14.   Diershke, E.G.; *All-Plastic Fiber-Optic Assemblies Simplify Design and Cut Costs;* Electronic Design 2, pp. 68-73, January 18, 1979.

15.   Bendiksen, L.F, and Intrieri, Jr., C.; *Fiber Optic Kits Simplify System Design;* Digital Design, pp. 36-41, May 1979.

16.   Ohr, S., and Adlerstein, S.; *Fiber is Growing Strong-Better Connectors, Cables Will Speed Things Up;* Electronic Design 23, pp. 42-52, Nov. 8, 1979.

CHAPTER 8

COMBINED OPTICAL AND HARDWIRE COUPLING

The combination of light sources, fibers and integrated optical devices which can be controlled by electrical integrated circuits offers several alternatives for the system and circuit designer of electro-optical systems.

8.1 INTRODUCTION

Integrated optics has been an active field of research for more than 10 years. During this time, considerable progress has been made toward realizing high-performance optical devices using guided-wave techniques such as optical switches and modulators. In addition to these devices, future single-mode optical communication systems will utilize wavelength filters to perform multiplexing/demultiplexing. Wavelength multiplexing allows better utilization of the very wide information bandwidth of single-mode fibers.

Various manufacturers have started producing multimate fiber optic ferrules that permit assembly of power, signal, coaxial, and fiber optic cables in a single standard connector housing. In this way, different cable types can be intermixed forming a combined hardwire that helps interconnections and interfaces in electro-optical systems.

8.2 ELECTRO-OPTICS AND ACOUSTO-OPTICS

The use of guided-wave techniques has led to switching devices with increased compactness, lower drive voltages and drive power compared to bulk electro-optical devices. Acoustic transformer devices combined with fiber optics can attain much higher isolation voltages than conventional transformers.

8.2.1 Electro-Optic Devices

An electro-optic device is an electrically controlled waveplate. That is, a material is employed that exhibits a change in index of refraction when an electric field is applied. The most common method employed is the Pockel's effect which is the linear electro-optic effect. There are two types of modulators that use the Pockel's effect: the transverse-field device and the longitudinal device.

In a transverse-field device, the halfwave voltage is directly proportional to the aspect ratio of the crystals. That is, the smaller the aperture for a given path length, the lower the voltage required to go from the off state to the on state. Figure 8.1 shows a dynamic transfer characteristic for an electro-optical modulator.[1] Typical applications include very broadband image reproduction systems, pulse selection of mode-locked lasers, high speed chopping, and noise-suppression systems.

The longitudinal Pockel's cell has its electrical field and optical path colinear. The halfwave voltage is dependent upon the number of crystals through which the laser beam passes. Since the power required to drive those devices over a broad bandwidth is prohibitively high, applications for this device are usually limited to very low repetition rates with very fast risetimes. Typical uses include Q-switches, chopping of CW lasers and pulse selection from a mode-locked train.

8.2.2 Acousto-Optic Devices

Acousto-optic devices have the capability of altering the amplitude, position, frequency, and polarization of a laser beam. An ultrasonic wave is launched into an acoustic-wave-supporting material by means of a piezoelectric transducer. As shown in Fig. 8.2, a laser beam is aligned at a specific angle, usually referred to as the Bragg angle.[1] Such devices are called Bragg cells. The parameters that determine the performance of a Bragg cell are : carrier frequency (transducer resonant frequency), acoustic velocity in the material

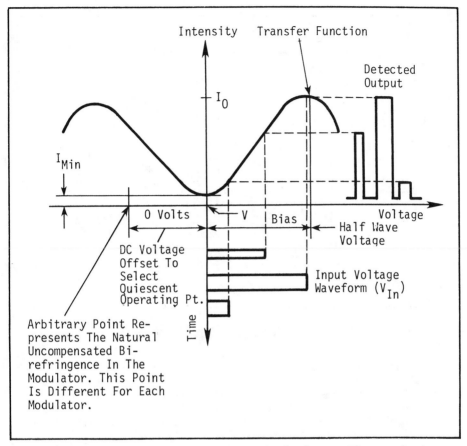

Figure 8.1 - Dynamic Transfer Characteristic for an Electro-Optic Modulator.

and laser beam diameter.[2] Recent research in planar guided-wave acousto-optics has made possible realization of high performance, thin-film Bragg modulators and deflectors with GHz bandwidth in Y-cut lithium niobate waveguides.

8.2.3 Voltage Isolation by Acoustics and Fiber Optics

It is possible to increase the usual 5-KV breakdown limit of isolation amplifiers to 40 KV or more using piezoelectric acoustic resonators and a fiber optic link. The key to the technique is a device called an *acoustic transformer* that transmits power along a glass or ceramic rod and isolates the power supply from the signal input.

8.3

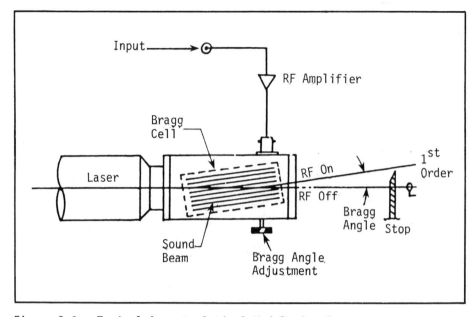

Figure 8.2 - Typical Acousto-Optical Modulation System.

Isolation amplifiers find use wherever a designer wants to electrically separate the inputs of a device from high common-mode voltages in it. For example, isolation protects patients from electrocution through the input electrodes of electrocardiograph machines. Another application of these amplifiers is interrupting ground loops between transducers and industrial data-collection systems.

In general, all isolation amplifiers have two critical design factors: (1) isolation of the power supply from the input and (2) galvanic separation of the signal line from the input terminals. The first factor is usually the more difficult problem of the two to solve. Conventional isolation amplifiers (Fig. 8.3) use a transformer to galvanically separate the power supply from the input, but this transformer becomes bulky and expensive as the voltages to be protected against rise beyond about 5 KV. The technique that uses an acoustic transformer (Fig. 8.4) approaches the problem of power supply isolation differently. The device uses an oscillator-driven piezoacoustic resonator at one end of a glass or ceramic rod to send acoustical energy along the rod at about 800 kHz. At the other end, an identical resonator receives the acoustical energy and converts it back to electrical energy. After rectification and filtering, the electrical energy powers the isolation amplifier. Such setups have achieved breakdown strength as high as 40 KV.

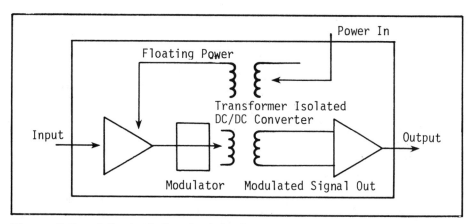

Figure 8.3 - A Typical Standard Isolation Amplifier.

As mentioned above, galvanic separation of the signal line from the input terminals constitutes the other critical isolation-amplifier design factor. Conventional devices modulate the signal across a transformer or optical isolator. Here again, transformers become bulky and optical isolators are good only up to a certain voltage level. Fiber optics solve the signal-isolation problem in Fig. 8.4 by stretching apart the two halves of the optical isolator system. The device works like a conventional optical isolator, but the distance between transmitter and receiver isolates the higher voltages.

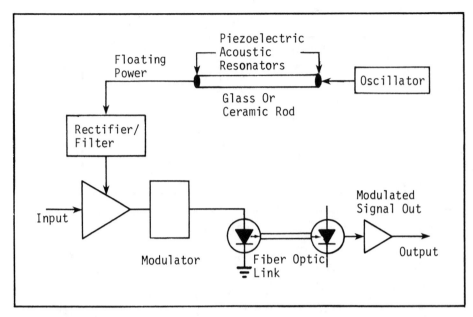

Figure 8.4 - An Improved Method in Isolation-Amplifier Design Uses a Fiber-Optic Link and a Glass of Ceramic Rod as an Acoustic Transformer to Separate the Input Line from the Rest of the Unit.

8.3 OPTICAL SWITCHING

Fiber optic switches are the equivalent of electrical multicontact switches and constitute a dependable means for routing signals from one path into one of two or more. The technology that contributes to construction of such devices includes at least three categories: (1) electro-optic, (2) magneto-optic and (3) acousto-optic.

8.3.1 Photoelectric Switches

One of the areas where fiber optics is finding wide-spread use is in the field of photoelectric switches.[4] The fibers can collect and distribute the energy into areas and under conditions where the use of photoelectric switches would not be practical. One useful configuration is a bifurcated bundle which has two branches terminated in a single probe end. One branch of the bifurcated bundle is connected to a photodetector, while the other is connected to a light source (Fig. 8.5). The common probe end can be used as a proximity sensor. When used as a photoelectric proximity sensor, the light carried to the sensing end by the transmitting fibers is reflected from the object being sensed, and the reflected energy passes through the receiver photodiode.

In through-beam applications, two fiber optic bundles are used, one to the detector and one to the emitter. The ends of the fiber optic cable are set up opposite each other and objects are sensed when the beam is broken. In the retroreflective mode, the light is beamed to a larger retroreflector and then back to the detector, via a bundle in which emitter and receiver fibers are combined, as in reflective sensing. When using retroreflective sensing, special care must be taken if the object being sensed is highly reflective or in close proximity to the sensing head. Recent improvements in optoelectronic devices make photoelectric sensors powerful enough to be able to operate over ever-increasing distances and even penetrate certain materials, such as plastics, glass, and fiberglass. A through-beam system can be set up to look for opaque parts inside translucent assemblies as they pass on a production line.

Fiber optics can add logic functions such as OR and AND. For example, it is possible to obtain OR logic (output from any optical fiber bundle individually activates the photoelectric switch) by alighing bifurcated cables with the common end facing the region to be sensed as shown in Fig. 8.6. One branch of each bifurcated bundle is connected to a common light source; the remaining branches are used as an input to a common photoelectric switch. Reflection caused by an object moving into the field being sensed and picked up by any or all of the fiber optic bundles, causes the photoelectric

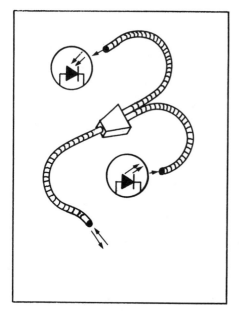

Figure 8.5 - Typical Bifurcated
Fiber Optic Bundle, where output
Jacket is Armored for Industrial
or other Critical Environments.

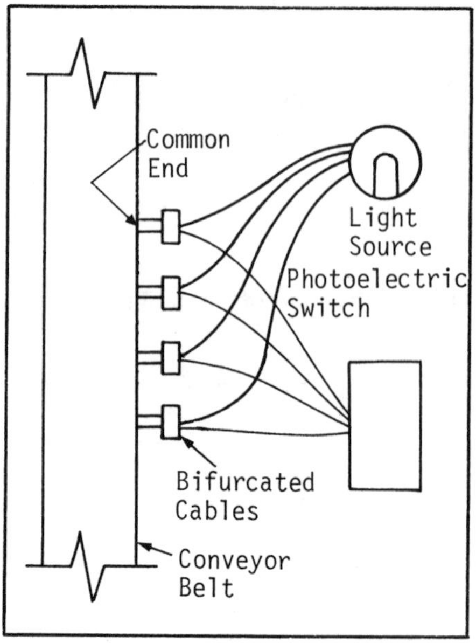

Figure 8.6 - Set-up that Implements
OR Logic Function. Any Sensor, Receiving a Reflection, Provides Sufficient Light to Actuate the Photoelectric Switch.

switch to activate AND logic (output from all sensors is necessary to
activate the photoelectric switch) is implemented in a different fashion. Several fiber optic cables are connected to a common light
source and several other cables (through-beam mode) are connected to
a common photoelectric switch. The ends of the cables from the emitter are aligned across the area being sensed with the ends of the
fiber optic cables connected to the photoelectric switch. The photoelectric switch, performing AND function, is adjusted to operate in
the absence of light. All of the optical paths must be blocked before the photoelectric switch will operate.

8.3.2 Liquid-Crystal Switch

An electro-optical switch based on liquid crystals[5] is shown in
Fig. 8.7. Here, four multimode fibers are coupled by rod lenses to
two glass prisms that in turn sandwich a 6-μm film of nematic liquid
crystal. An external electric field controls the switching by rea-
ligning the liquid's molecular order. The 1.6 refractive index of

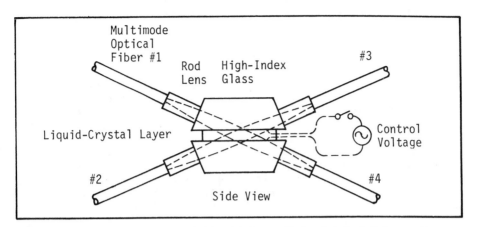

Figure 8.7 - A 6-μm Layer of Liquid Crystal Sandwiched Between Two Trapezoidal Prisms Forms the Heart of the Switch.

the glass is so high that the liquid acts either as a lower-index cladding on the glass or as a leaky dielectric for light obliquely incident on the glass-to-liquid interface. Light from the input fiber is either partially or totally reflected at the interface, depending on the order of the liquid. This is determined by the control voltage.

Liquid-crystal switches have shown insertion losses as low as 1.6 dB with no voltage. Table 8.1 shows the characteristics of electro-optic liquid crystal switches[6] along with the characteristics of other switching techniques.

Table 8.1 - Comparison of Multimode Fiber Optic
Switching Characteristics (*After Ref. 8*).

	Electro-mechanical	Electro-optic LiTaO3	Electro-Opt. Liq. Crystal	Magnetic-Optic
Optical Insertion Loss	1dB (approx)	5.7 to 12dB	1dB (approx)	4dB minimum (theoretically)
Switching Speed	ms	ns	ms	μs
Optical Crosstalk	-40 dB	-26-to-12dB	-48 dB	-40dB est
Reliability	moving parts	high	high	high
Control Power Requirements	2 to 4V 0.1 to 0.2A	400 to 500V μA to mA	5 to 30V 20 μA	2 to 4V 0.1 to 1.0A

8.3.3 Optically Coupled Microwave Switches

It is possible to use lightwaves to couple microwave energy between two points via a switch with direct modulation of laser diodes.[7]
The basic configuration for the switch is shown in Fig. 8.8.   It consists of an input section containing a laser (as an optical source)
and an output section containing a p-i-n or avalanche photodiode (as
an optical detector). The two sections are isolated from each other
at RF frequencies by proper shielding. In the input section, the RF
signal modulates the intensity of the lightwave emitted by the optical
source at the RF frequency. The modulated lightwave is then coupled
to the optical detector where it is demodulated. The bias control
signal determines the detector's sensitivity, thereby permitting continuously variable attenuation or on/off switching.

There are three important features of this switching concept.
First, the reverse isolation (the isolation of the input from signals
present at the output) is virtually infinite provided that either photon emission by the detector or photon detection by the source is
negligible. Second, the on/off ratio of this device can also be

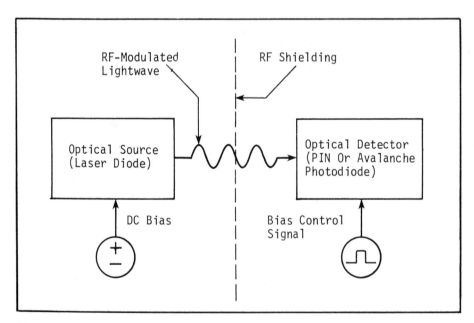

Figure 8.8 - Basic Configuration of an Optically Coupled Microwave
Switch (*After Reference 7*).

extremely high since it is ultimately limited only by the degree to which the detector can be turned off. Third, as a result of the virtual infinite reverse isolation, the input impedance of the switch is independent of its state, and switching transients are prevented from reaching the input.

The characteristics of this device make it particularly attractive for a variety of radar applications, where high on/off ratio and high reverse isolation improve sensitivity and jamming immunity. In addition, the fact that the input impedance of this switch is independent of the switching state makes it attractive for applications where oscillator *pulling* must be avoided.

8.3.4 Electro-Optic Tunable Filters

An Electro-Optic Tunable Filter (EOTF) makes the tuning of optical filters easier by giving independent control of operating wavelength, linewidth, line shape and peak transmission. Such an EOTF has been developed by Hughes and operates at dc rather than RF.[8,9] This device (Fig. 8.9) uses much less drive power than similar acousto-optic interaction devices (AOIs). A further advantage of the EOTF is that its transfer function can be synthesized under microprocessor control. Hence, to synthesize such features as sidelobe suppression, multiple-wavelength operation and variable linewidth, it is necessary only to program the microprocessor (μP).

The EOTF device shown in Fig. 8.9 consists of a 100-μm thick platelet of $LiTaO_3$ with 100 transverse electrodes per surface space, approximately 50 μm apart. Separate driver-circuits apply the required voltage to each electrode pair. These drivers are, in turn, controlled by the μP. To change the filter's transmission, the processor first computes the spatial period that corresponds to the crystal's birefringence and to the wavelength required after coupling. Then the μP develops each driver's voltage by sampling the resulting distribution of voltages. After the voltages are computed and stored, switching the filter takes 2 ms, using serially addressed drives.

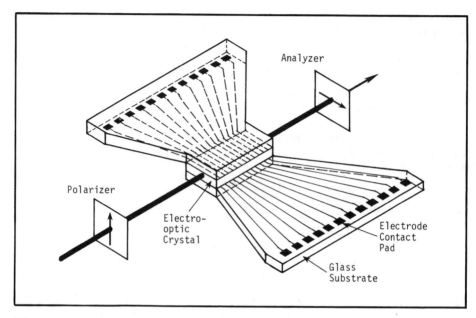

Figure 8.9 - Electro-Optic Filter for the Visible Spectrum Developed by Hughes Researchers.

8.4 MIXED HARDWARE AND FIBER OPTIC MULTIPLEXING

A single optical fiber can carry a large volume of communication traffic - video signals for several subscribers plus telephone service to each, with additional capacity for data and control signal transmissions. If one fiber can be made to serve more than one subscriber, the transmission system cost is reduced. An efficient method is therefore needed to combine the signals for several subscribers into a common fiber for transmission and then separate them at the far end for delivery. One approach is to use state-of-art applications of Wave Division Multiplexing (WDM) technology, that is, techniques for combining and dividing light signals by means of their wavelengths.

8.4.1 Hybrid Optical System With Keyboard-Entered Source Data

Figure 8.10 shows a block diagram of an optical data link, where N-computer aided data entry keystations are connected to a processor at one end via coaxial cable and N-keystations at the remote end, connected again via coaxial cable. An optical cable, with the two end optical modems that have replaced an equally long coaxial cable and the respective conventional modems, make this arrangement an optical system with hybrid or mixed hardware. This data link for computer-aided data entry is assumed to operate in a half-duplex mode. Hence, data terminals can be interconnected by a single fiber cable with a tee-coupler at each end of the optical cable.

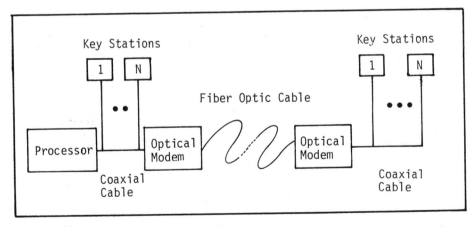

Figure 8.10 - Hybrid Optical System with Keyboard-entered Source Data.

In a real system of similar structure, as reported by Goldberg and Eiber,[10] keyboard-entered source data from a maximum of 32 operator key stations produce validated data records for mainframe processing. Systems requirements include a data link length up to 1.6 km, operation at 10^6 baud, a bit error rate less than 10^{-15}, and half-duplex data transmission. In this case the optical cable passes through manholes, under a bridge, through a subway tunnel and over-head structures.

A similar structure could be used as a trunk or spanline between two telephone central offices, including multiple fiber optic cables, duplex links and probably one or more repeaters for data transmission over longer distances. It is worthy of mention that Harry Diamond Laboratories in Maryland has put together 35 fiber optic cables total-ing 7.5 kilometers that are used as trunk lines to carry 16 multi-plexed, full duplex RS-232 channels from a computer mainframe to re-mote distribution points serving 256 peripherals.

8.4.2 Wavelength Division Multiplexing (WDM)

In conventional fiber optics, signals from different light sources require separate and uniquely assigned optical fibers. In contrast, WDM permits a more efficient utilization of the information-carrying capacity of the optical fiber.[11] WDM allows simultaneous transmission of optical signals from different light sources through the same optical fiber while preserving the message integrity of each optical signal for subsequent conversion to corresponding electrical signals.[12]

WDM techniques use LEDs or lasers as light sources, each source emitting at a different wavelength. To combine the light from each source to the same fiber, the WDM approach uses a device called an optical multiplexer (Fig. 8.11a).[13] Photodetectors at the far end of the fiber are broadband devices with respect to the wavelengths being emitted; that is, they respond to many different wavelengths but can-not distinguish one from another. Another device is therefore needed which can separate the light into its component wavelengths. Such a device is called an optical demultiplexer.

8.4.3 Wavelength Demultiplexing

Demultiplexing the multiple color signals on the cable requires some type of dispersion effect. For simplicity, a transmission grat-ing is used in Fig. 8.11b. The output beam from the fiber is convert-ed to a line source, using a fiber optic circular-to-linear converter. This line source is then imaged onto the transmission grating. A

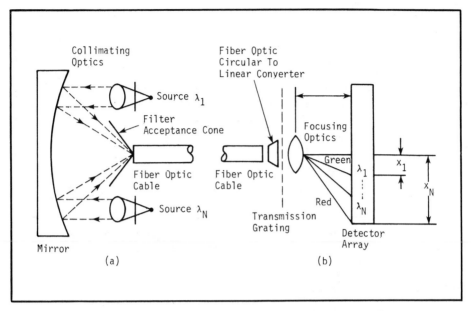

Figure 8.11 - Wavelength Multiplexing/Demultiplexing. a) Cross-section of Optical Multiplexer, b) Schematic Drawing of Demultiplexer.

lens on the output side of the grating focuses the colors onto the detectors. By choosing the proper combination of grating period and output lens focal length, the linear distance between the images of the individual colors can be controlled. As can be seen from Fig. 8.11b, the linear distance, x, between colors is equal to:

$$x = \frac{Pf\lambda}{d} \qquad (8.1)$$

where,   P = the order of dispersion (=1)

f = the output lens focal length

d = the grating spacing

λ = the wavelength.

Wavelength division multiplexing techniques can also be used for bidirectional transmission as shown in Fig. 8.12. Each multiplexing device operates in two directions. The first device acts as a multiplexer for the signal from light source 1 and as demultiplexer for the signal from light source 2. The second multiplexing device does the reverse. In this way, the signal from light source 1 travels in one direction down the fiber, while the signal from the other light

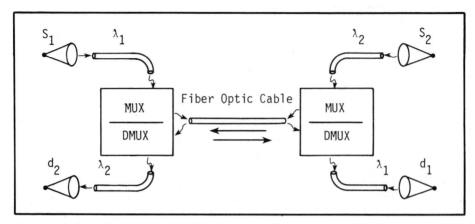

Figure 8.12 - Example of Bidirectional Transmission Through a Single Fiber Using a Combined Wave Multiplexing/Demultiplexing Method.

source travels up the fiber in the opposite direction.

        With the improvement of optoelectronic devices, the designers of fiber optic bus organized systems will be able to take advantage of the WDM techniques in the future, and apply them toward the design of computer communications systems.[14]

8.5 REFERENCES

1. Rizzo, R.; *Laser Beam Modulators: Electro-Optic or Acousto-Optic?* Electro-Optical Systems Design, pp. 37-41, October 1979.

2. Tsai, C.S.; *Guided-Wave Acoustooptic Bragg Modulators for Wideband Integrated Optic Communications and Signal Processing;* IEEE Transactions on Circuits and Systems, Vol. CAS-26, No. 12, pp. 1072-1098, December 1979.

3. Peterson, B.; *Acoustics, Fiber Optics Up Isolation Voltages;* EDN, p. 62, November 20, 1978.

4. Krueger, A.H.; *Applying Optics to Photoelectric Switches;* Control Engineering, pp. 61-62, August 1980.

5. Reference 27 of Chapter 4.

6. Adlerstein, S.; *3 Technologies Gearing Up for Optical Switch Race;* Electronic Design, p.36, October 11, 1979.

7. Kiehl, R.A. and Drury, D.M.; *An Optically Coupled Microwave Switch;* The IEEE MTT-S International Microwave Symposium, Washington, DC, May 28-30, 1980.

8. Adlerstein, S.; *EOTF Independently Controls Wavelength and Linewidth;* Electronic Design 19, pp. 62-64, September 13, 1979.

9. Schmidt, R.V. and Alferness, R.C.; *Directional Coupler Switches Modulators, and Filters Using Alternating $\Delta\beta$ Techniques;* IEEE Transactions on Circuits and Systems, Vol. CAS-26, No. 12, pp. 1099-1108, December 1979.

10. Golderber, N. and Eiber, J.A.; *Optical Link Design and Component Selection;* Computer Design, Vol. 18, No. 5, pp. 218-222, May 1979.

11. Straus, J.; *Wavelength Division Multiplexing (WDM);* Telesis, pp. 2-6, 1980 (Two).

12. Georgopoulos, C.J.; *Large Optically Coupled Systems With Combined Hardware;* Simulation of Distributed-Parameter and Large Scale Systems, pp. 33-37, IMACS, 1980.

13. Flanagan, M.; *Color Multiplexing on an Optical Fiber;* Electro-Optical Systems Design, pp. 31-36, March 1979.

14. Georgopoulos, C.J.; *The Fiber Optics in Computer Communications;* The 28th International Scientific Congress of Electronics, Palazzo dei Congressi, Rome-EUR, 23-24, March 1981.

CHAPTER 9

FIBER OPTICS IN CONTROL AND MEASUREMENTS

Optical fibers offer several advantages over copper wires when used in industrial processes and instrumentation, the most important being noise immunity, safety, security, size and weight.

9.1 INTRODUCTION

There has been an increasing need for the development of optical sensors where optical fibers are used as signal transmission lines. As optical fibers are used for the signal transmission, the electrical isolation between optical sensor and measuring instruments can be secured. Furthermore, optical fibers are so insensitive to electromagnetic interference that measurements in a high-voltage power system can be safely and accurately performed with optically-lined telemetry systems.

For non-destructive measurements of certain fiber optic system parameters, instruments based on optical time-domain reflectometry are used. As far as hardened design techniques are concerned, the manufacturers' effort is directed toward the use of proper dopants when making the fiber optic cables, while system designers place special emphasis on the hardening of the receiving end of a fiber optic link.

9.2 FIBER OPTICS IN INDUSTRIAL CONTROL

Conventional data acquisition systems hardware is an area where fiber optics will have a great effect. As distributed control becomes an accepted technique in processes, distributed data acquisition systems will be required to maintain close communication with central control for optimal cost effectiveness.[1] In this respect, sensing devices and fiber optic cables are extremely useful, particularly in regions where explosives or other hazards may exist.

9.2.1 Fiber Optics for Sensing Process Variables

To exploit the fiber optic technology for industrial control applications, optical sensors have been developed to measure pressures, temperatures, and positions (both rotary and linear). Ideally, these sensors should interface directly with optical fibers.

In general, a sensor is located at a point of measurement and a fiber optic cable transmits an optical signal to it. The sensor modulates the optical signal, in some fashion, and sends it to an optical detector over another optical cable. In many applications, the optical sensor and fiber optic cable must survive extremes of temperature and vibration. The light source and light detectors usually are remotely located in a controlled environment suitable for electronic hardware items.

Two representative types of sensor applications using fiber optic technology are shown in Fig. 9.1.[2] The first circuit (Fig. 9.1a) represents a pressure sensor. As the diaphragm moves in response to changes in pressure, the intensity of the light reflected off the diaphragm is proportional to these pressure changes. The second circuit (Fig. 9.1b) shows a rare-earth temperature sensor developed by United Technologies Research Center for NASA Lewis.

9.2.2 Intraplant Link for Process Control

An intraplant link for process control data systems may be a fiber optic link that ties a central computer to a distributed microcontrol computer via TTL logic I/O. In this case, the input and output must be fully buffered in order to be used with digitized data channels. Other aspects include interface circuitry, data formats, protocol functions, timing, and software-related issues.

Interfaces have been developed that extend the standard IEEE 488 capabilities for distribution or collection of information to and

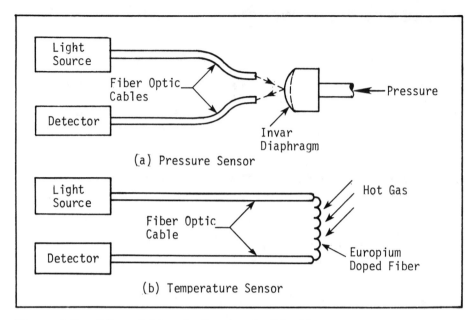

Figure 9.1 - Fiber Optics in Sensing Applications.

from remote areas around a laboratory or industrial process.   One
such interface is the 12050A fiber optic HP-IB link[3] (developed by
Hewlett-Packard).   Figure 9.2 shows a typical use in a production
environment and demonstrates the ease of duplicating the test set-up
while maintaining the advantages of single-computer control.

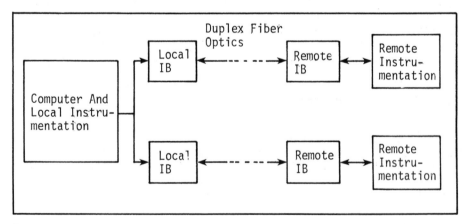

Figure 9.2 - The Fiber Optic Link Makes it easy to Duplicate a Group
of IB Instruments at Several Sites, Using a Single Computer at one
Site.

9.3 MEASURING TRANSMISSION CHARACTERISTICS OF FIBER OPTICS

Fiber optics measurement technology is still in the development stage, with current instrumentation evaluating only a few major parameters, such as fiber/cable attenuation, quality, bandwidth and light power.

9.3.1 Basic Optical dB Units[4]

Measurements of optical radiation power are expressed in watts. Decibel (dB) power is:

$$dB = 10 \log \left[\frac{P_{sig}}{P_{ref}} \right] , \qquad (9.1)$$

where, P_{sig} is the power to be measured and P_{ref} is the reference power. For 1-mW reference power:

$$dB_m = 10 \log \left[\frac{P_{sig}}{1 \text{ mW}} \right] \qquad (9.2)$$

For 1-µW reference power:

$$dB_\mu = 10 \log \left[\frac{P_{sig}}{1 \text{ µW}} \right] \qquad (9.3)$$

With both P_{sig} and P_{ref} variable, the dB-power formula expresses the log ratio of the two unknowns in dB.

Light-power loss of an optical data-link element is:

$$L(dB) = 10 \log \left[\frac{P_o}{P_{in}} \right] \qquad (9.4)$$

The input power (P_{in}) and output power (P_o) of a component can be measured in dB_m units, dB_μ units, or dB units without a known reference. The loss, L, expressed in decibels, is numerically the same:

$$L(dB) = dB_m \left(P_o \right) - dB_m \left(P_{in} \right)$$

$$= dB_\mu \left(P_o \right) - dB_\mu \left(P_{in} \right) \qquad (9.5)$$

9.3.2 Optical Time-Domain Reflectometer[5,6,7]

If both ends of a cable are available, a relatively unsophisti-
cated equipment may be used to measure continuity, attenuation, bandwidth and other parameters. With only one cable end available, more
complicated equipment, such as optical time-domain reflectometers
(OTDRs) will be required.

An optical time-domain reflectometer utilizes backscattering and
far-end reflection effects in an optical fiber to determine the fiber's length, the location of breaks, the nature of splice and connector losses, the amount of fiber attenuation and the status of any
manufacturing defects. It makes these measurements using only one
end of the fiber's cable.

An OTDR closely resembles the time-domain reflectometer employed
in coaxial-cable measurements.

9.3.3 Fiber Optic Multiconnection Testing Set-Up

Systems with mixed hardware, when installed, need special arrange-
ments for testing. For fiber optic links, special attention must be
given to the optical characteristics of the cable which can be affected by the physical layout, joints, or localized optical dishomogene-
ities. The design of the testing system in this case is rather complicated because of the transmission design mix of fiber optic and
coaxial technology in a single system.

A testing system by which both coaxial cable and optical fiber
components can be tested is shown in Fig. 9.3. This integrated system is equipped with measurement software, which allows control of
the hardware and development of complex measurement algorithms with
flexibility. It combines a digital processing oscilloscope (DPO)
with the 7S12 sampler. Similar testing systems have been described
in various papers (Ref. 6 and 8), where time measurements with reso-
lutions of tenths of picoseconds and frequency measurements in the
range from DC to a few GHz have been reported. Parameters that can
be measured include attenuation of long cables, spectral attenuation
and pulse dispersion from which cable bandwidth, phase delay, and
fiber transfer function are calculated. The system also is an optical TDR, which is useful in locating fiber breaks, determining fiber
length, and evaluating connector and splice losses. In the case of
coaxial cables, slotted lines, i.e., precision lengths of coaxial
cable with a slit in their outer shielding conductor allowing access
to the probe, are used to measure VSWR. The digitizing oscilloscope
collects VSWR data by utilizing a slotted line connected to the
length of coaxial cable under test. To evaluate optical-cable

parameters using pulse-spreading data, an input pulse on a reference length of fiber is first measured, and then the spreading of the same pulse on a longer length of fiber optic cable is detected.

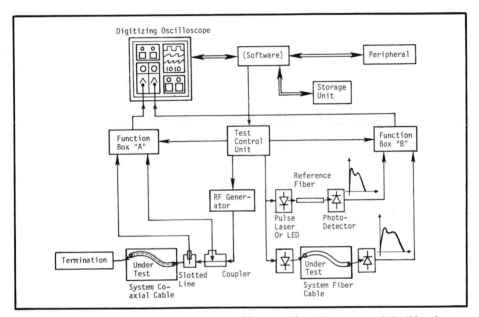

Figure 9.3 - Test Set-up for Optically Coupled Systems with Mixed Hardware.

9.4 OPTICAL LINKS IN EMI ENVIRONMENTS

Fiber Optics have already been adopted by electrical power companies owing to their immunity to electromagnetic interference and to their capability of high speed transmission of information.

9.4.1 Electromagnetic-Field Hazard Probes

Conventional hazard probes, with metallic output wires, scatter and reflect the electromagnetic field in the vicinity[9] of the probe, obscuring the true field distribution. Other field measuring systems, utilizing high-resistance output cables to reduce such interaction, introduce increased background noise.

The above problems can be alleviated using an optically linked system. A typical optically linked telemetry system (Fig. 9.4) accepts low-level analog input signals (dc to 2 kHz, 0-50 mV peak) via a high-input-impedance micropower differential amplifier.[10] The conditioned analog signal is fed into a voltage controlled-oscillator (VCO) which produces a square wave output pulse-train whose frequency (15-30 kHz) is linearly proportional to the analog input signal. The VCO output drives an LED diode and the fiber-optic link optically couples the telemetry transmitter's output to the photodiode detector(s) in a receiving system which can be located out of the high-field-strength zone. In general, radio-frequency interference, which occurs in several hardwired instrumentation systems placed in strong RF fields (below 1000 MHz), is significantly reduced through the substitution of the optically-linked telemetry system.

9.4.2 Use of Fiber Optics in the Power Industry

Fiber optics can be used in the power industry at almost all levels. Fiber optic cables are available to fit several needs, i.e., duct, aerial, or direct burial. As cable costs drop and longer wavelength devices become available, there will be an increasing utilization of fiber optics in the power industry in applications such as the following:

- Communications between buildings.

- Entrance links to power stations.

- Interface in measuring set-ups.

- Power control in inter-city systems.

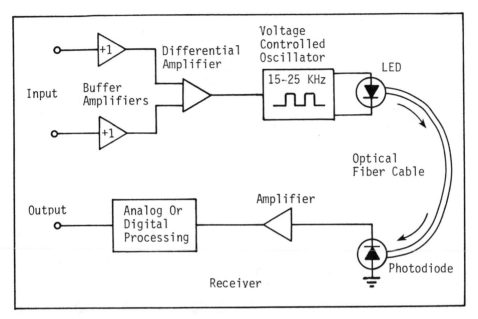

Figure 9.4 - Typical Optically-Linked Telemetry System.

- Data acquisition and alarm networks.

- Miniature indicators.

    When selecting fiber optics for power utilities, several pieces of equipment must also be considered for installation and maintenance. Such equipment may include:

- Optical power meter,

- Optical time-domain reflectometer,

- Optical fiber cutting and splicing devices.

9.5 HARDENED FIBER OPTIC LINKS

Hardened design techniques have been developed for fiber optic links that consider transient ionization and permanent damage effects. System requirements usually dictate the bit rate, bit error rate (BER), link recovery time, and fiber length. In an unhardened link, the most critical factors that influence the choice of wavelength and receiver sensitivity are the coupled optical power, fiber loss, and fiber dispersion.

9.5.1 General Considerations

In a hardened fiber optic system, the performance of the fiber and emitter is very important, but the properties of the detector and receiver play a much larger role. The selection of wavelength and receiver optical sensitivity must consider the effects of radiation on all of the link components. The final configuration requires a trade-off of the factors[11] listed in Table 9.1. For example, unless the transient radiation environment is extremely low, radiation-hardened receivers cannot use avalanche photodiode detectors which provide better optical sensitivity than PIN detectors and are frequently used in unhardened applications. For this reason, hardened receivers generally have lower optical sensitivity than conventional unhardened receivers. Radiation-induced absorption losses in fibers are a strong function of wavelength, fiber type, and fiber length.

Table 9.1 - Fiber Optic Data Link Design Parameters

SYSTEM-DEFINED PARAMETERS	STANDARD DESIGN PARAMETERS
Link Length	Wavelength
Data Bit Rate (Bandwidth)	Emitter Output Power
Bit Error Rate	Coupled Power
Down-Time Allowed	Intrinsic Loss in Fiber
Temperature Range	Fiber Dispersion
Radiation Environment	Detector Responsivity
	Receiver Sensitivity

RADIATION-DEPENDENT PARAMETERS

Emitter Degradation
Induced Loss in Fiber
Fiber Recovery Rate
Induced Noise in Detector
Detector Degradation

9.5.2 Fiber Optic Cable Hardening

The problems caused by radiation are serious. High doses of radiation can induce attenuations in optical fibers far in excess of their nominal attenuation. For example, a fiber manufactured with an attenuation of a few dB/Km in the IR spectrum could be radiation-sensitive enough to suffer thousands of dB loss through exposure (Fig. 9.5).[12]

Presently, research, testing and other related investigations are being conducted by various laboratories to discover methods for fiber optic hardening. The main effort is directed toward the use of proper dopants, with germanium and boron being the basic ones. However, it has been observed that phosphorus, as a dopant, almost totally suppresses the so-called transient upset. If, for example, a fiber with no phosphorus is hit with a radiation pulse such as that produced by a nuclear weapon, losses as high as 20,000 to 30,000 dB would occur for short periods. When phosphorus is introduced as a dopant, however, the damage decreases many orders of magnitude.[12]

Special tests have been conducted on germanium-silicate fibers doped with boron, antimony, phosphorous, and/or cerium for their responses to steady-state and transient ionizing radiation.[13] Results have indicated that the addition of antimony to germanium-phosphosilicate fiber improves its response to steady-state radiation without reducing its favorable transient response. The addition of cerium to germanium-borosilicate fiber can improve its response in high-dose environments.

In general, for radiation environments with short links, polymer fibers are good, provided other factors, such as the temperature environment, are acceptable. At about 1 to 1.1 μm, the radiation sensitivity of most silica fibers is minimal. For the long-range view, by coincidence, this favors the current trend towards the development of LEDs and lasers operating in these long-wavelength regions.[12]

9.5.3 Receiver Hardening Concepts

There are several different approaches that can be used to harden receivers, including the use of a dual optical channel differential system, consideration of modulation and encoding methods, and the trade-off between sensitivity and transient ionization hardness.[11] A dual optical link provides improved γ-hardness because fiber fluorescence and common-mode photocurrents are rejected by the differential system. If fiber fluorescence is not important, a single optical channel can be used in a differential receiver to provide first-order cancellation of detector photocurrents. Although more optical

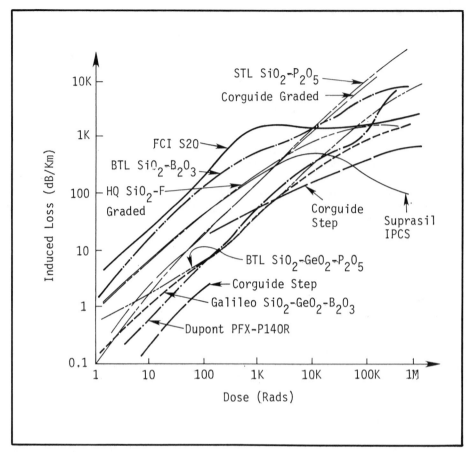

Figure 9.5 - Radiation-Induced Optical Attenuation.

components are required, the γ-hardness of the receiver can be improved by a factor of 10 or more with the differential technique.
There are also advantages in considering different types of modulation methods for hardened optical link designs. Although a simple
pulse code modulation (PCM) approach is the simplest implementation,
the required S/N ratios are lower for frequency-shift keyed (FSK) or
phase-shift keyed (PSK) modulation methods.

9.6 REFERENCES

1. Bailey, S.J.; *Fiber Optic Data Links Near for Industrial Control;* Control Engineering, pp. 32-35, December 1978.

2. Baumbick, R.J. and Alexander, J.; *Fiber Sense Process Variables;* Control Engineering, pp. 75-77, March 1980.

3. Crady, R.B.; *High-Speed Fiber Optic Link Provides Reliable Real-Time HP-IB Extension;* Hewlett-Packard Journal, pp. 3-4, December 1979.

4. Wendland, P.; *Lighten the Burden of Fiber-Optic Measurements with New Instruments, Standards;* Electronic Design 21, pp. 126-129, October 11, 1979.

5. Andreiev, N.; *Newest Fiber-Optics Test Instruments Shed Light on Burgeoning Applications;* EDN, pp. 45-50, November 1978.

6. Ramirez, R.W.; *Digitizing-Oscilloscope Systems Simplify Transmission Measurements;* EDN, pp. 117-123, June 5, 1979.

7. Rourke, M.D.; *An Overview of Optical Time Domain Reflectometry;* American Ceramic Society, 82nd Annual Meeting, Chicago, Ill., April 27-30, 1980.

8. Reference 12 of Chapter 8.

9. Bassen, H.; *A Broad-Band Miniature Isotropic Electric Field Measurement System;* 1975 IEEE Electromagnetic Compatibility Symposium Record, IEEE Publ. 75 CH1002-5 EMC, pp. 5BIIa (1-5), October 1975.

10. Bassen, H.I. and Hoss, R.J.; *An Optically Linked Telemetry System for Use with Electromagnetic-Field Measurement Probes;* IEEE Transactions on Electromagnetic Compatibility, Vol. EMC-20, No. 4, pp. 483-488, November 1978.

11. Johnston, A.H. and Caldwell, R.S.; *Design Techniques for Hardened Fiber Optic Receivers;* IEEE Transactions on Nuclear Science, Vol. NS-27, No. 6, December 1980.

12. McDermott, J.; *Fiber Optics: Military Interest Spurs Advances On All Fronts;* Electronic Design, pp. 75-78, October 25, 1978.

13. Wall, J.A. and Posen, H.; *Radiation Hardening of Optical Fibers Using Multidopants Sb/P/Ce;* Conference on Physics of Fiber Optics, Chicago, Ill., April 28, 1980.

CHAPTER 10

ASSESSMENT OF CURRENT CAPABILITIES

This chapter discusses current work and capabilities of optical technology and project costs compared to current values and expected future developments.

10.1 INTRODUCTION

Fiber optic components and systems are evolving faster than projected and developing countries embarking on new networks offer greater scope for the mass penetration of fiber optic communication systems.

Data rates of 400 Mbits/s (Japan) and 322 Mbits/s (Canada) have been achieved in systems that have been installed and are operational.[1,2] At data rates above 100 Mbits/s, achievable cable bandwidths dominate system design constraints since, at these data rates, systems are more limited by dispersion than by attenuation. For systems using data rates below 100 Mbits/s, cable attenuation is the most significant parameter limiting span length.

In certain cases, cost considerations have weighed heavily against performance. Manufacturability and price difficulties have cropped up especially in the source-to-fiber and detector-to-fiber connections.[1] ,However, it is foreseen that within the decade of the 80's, optical fiber communications systems production levels will increase to the point where their installation may be justified with alternate transmission media on a strictly cost-competitive basis. A continuing chain of developments in this technology will spur an increase in production levels, which should result in further price decreases.

10.2 STATUS OF RESEARCH AND TECHNOLOGY

It can be stated that significant progress has been made in all
the key optical components: sources, detectors and fibers. However,
as optical fibers gain general acceptance and are considered for an
ever-increasing range of application, the necessity of basic research
into material systems and fabrication processes has re-emerged.

10.2.1 Progress in Light Sources

In the area of light sources, the prevalent devices are LED and
injection-laser diodes, which emit in the 0.85 µm wavelength range.
While LEDs are cheaper and last longer - with 100,000 hours of operation already obtained and 1 million hours certain to come- they only
can launch 1 to 2 mW into a fiber and can operate at only 100 Mbits/s.
Presently, LEDs are being investigated that can behave as both emitters and detectors. Such devices could enable two-way communications
to pass over a single fiber.

Laser diodes can launch 5 to 10 mW and operate at over 1 Gbit/s;
their lifetimes are continuously improving. Laser diodes have already operated in the field for over 10,000 hours without significant
deterioration. For longer wavelength operation (1.1 to 1.6 µm range),
stripe-geometry emitters will be the answer.[2] They will consist of
such ternary materials as gallium arsenide as well as several quaternary materials.

10.2.2 Progress in Light Detectors

For many applications in the optical communications field, avalanche photodiodes are the preferred detectors. This stems from the
tradeoffs involved in repeater spacing, the behavior of signal-to-
noise ratios, and the signal amplification property of the detector.
What is generally true for silicon APDs at 0.9 µm is just as true for
other APDs near 1.3 µm. At present, Ge APDs are the only commercially available detectors capable of filling the requirements of these
longer wavelength systems. However, Ge APDs suffer from high noise
currents, large temperature coefficients, and high voltage requirements. Therefore, the search is towards high-quality APDs made from
InGaAsP or GaAlAsSb alloys.

10.2.3 Progress in Fiber Optic Cables

Presently, graded-index fibers with doped-glass-core diameters of 50 to 65 μm carrying 0.8 to 0.9 μm radiation, present MHz·Km products in the hundreds. At longer wavelengths (1.1 to 1.6 μm), these products can be ten times greater. Without repeaters, Nippon Tel. & Tel. Public Corp. (Tokyo, Japan) has transmitted 100 Mbit/s signals over 52.6 Km of graded-index fiber using 1.27 μm radiation. On the other hand, small-core single-mode fibers, with 2 to 5 μm diameter cores, can have several GHz·Km products.[2] Their tiny cores make launching any appreciable light into them a problem outside the laboratory. Fusion in a gaseous arc is becoming the preferred method when telecomm. fibers are to be permanently joined (spliced). Fusion splices have at least half the strength of virgin fiber and only 0.2 dB loss.

Bell Laboratories' researchers have devised and patented a method to increase the rate at which fiber optics can be fabricated. The patent was granted in August 1980. The new technique is a variation of the generally practiced method of the Modified Chemical Vapor Deposition (MCVD) process.[3] The revised MCVD process uses a plasma, a highly energetic mixture containing ionized atomic fragments, to speed up preform production, an early step in the fabrication process. This process results in acceptable transmission losses and is expected to make fiber optics more economically competitive with alternative transmission media. Transmission losses are only 3.4 dB/Km at 0.85 μm and 1.5 dB/Km at 1.2 μm, compared to losses with present techniques.

10.2.4 Research on New Materials

The Japanese are investigating new types of materials for optical communication fibers in the 4-5 μm wavelength band, specifically metal halides. Theoretical predictions foresee an eventual attenuation loss of only 0.01 dB/Km.

The discovery of new materials and new possibilities has led to the need for studying these materials carefully in the laboratory to determine their properties in comparison with materials in current use. Thus, as fibers are moving into the phase of practical usage, the trend of research starts a second round, beginning again at examining the fundamentals.

10.3 SOME WORLD-WIDE EFFORTS AND APPLICATIONS

Fiber optic systems are moving beyond the experimental stage into various applications throughout the world. The technology and systems being deployed in Europe are similar to those used in the United States and Japan for compatible applications. The overall trend is an exponentially increasing variety and number of applications as the advantages of using fiber optics for these applications become more widely recognized.

10.3.1 General Competitive Status

According to the Commerce Department Survey of the U.S. fiber optics industry, presented at the 1980 Fiber Optic Conference by Joe Hull, the relative competitiveness of countries in fiber optics appears as shown in Table 10.1.

Table 10.1 - Competition for International Trade

ASSESSMENT DIMENSION	COUNTRY			
	USA	JAPAN	EUROPE	CANADA
Technology Rank*	2	1	3	4
Applications Development	4	1	3	2
Field Trial/ Installation Activity	2	1	3	4

* 1 = highest, 4 = lowest.

10.3.2 United States of America

In the USA, a considerable number of experimental, preoperational and operational systems have been planned or realized.[4] GTE, as a telephone company and manufacturer, has initiated a number of field trials. The latest GTE system, planned for Washington state, will ultimately provide 3360 voice channels over 9.6 Km. A ten-fiber cable, supplied by General Cable Corporation, will have a loss of 3.5 dB/Km and a bandwidth of 400 MHz/Km.

The Bell System has fiber optic links installed or planned in 12 locations to the middle of 1981. Nearly 12000 Kms of fiber (as distinct from cable) will be used. Other independent telephone companies are also active in various applications.

10.3.3 Japan

Progress in low-loss optical-fiber technology, including both device fabrication and transmission equipment, is so advanced in Japan that such systems are considered practical for several commercial applications. Much of the advanced development work has focused on wavelength division-multiplexing technology: high bit-rate system design; small-dispersion wavelength devices; low-loss laser diode light sources; high-speed modulation; single-mode laser diodes; and analog modulation laser diodes.[5]

Industrial fields for applications in Japan include steel production, power generation, petrochemicals, etc. At this time, Japan is pushing research and development and beginning to cultivate overseas markets. Since April 1981, the government corporation has been using fiber optics in its public telephone networks in the Tokyo, Chiba and Kawasaki areas. Based on the results of the short-distance service, the communications company is already considering expanding the new technology to a nationwide system.

10.3.4 Europe

In Europe there is interest in the use of fiber optic links for wideband services to customers. Such services range from business teleconference links, involving quality video and audio two-way transmission, to interactive community service (CATV) trials. The teleconference links are probably nearest to being realized, but there are very wide variations in the wideband services from country to country in Europe.

Figure 10.1 summarizes the first generation of European optical fiber systems (1977-1979) according to J. Midwinter of the British Post Office.[6] Additional applications are occurring in a wide range of industries such as power, medicine, nuclear energy, military and many others.

10.5

10.3.5 Canada

The trend in Canada is generally towards less restrictive division of responsibility of carriers, and field trials are progressing in a number of areas. Among other activities, Bell Canada has started a field trial of a fiber optic system for home telephones. The program includes the test of a voice, video, and data service to 35 homes in Toronto. The installation has 1.2 kilometers of graded-index fiber optic feeder cable, plus about 200 meters of graded-index entrance cable for each of the subscribers.

While most installations in North America deal only with links from one central office to another at 44.7 Mb/s, the Canadian installations are used in two-way subscriber loops to telephone customers.

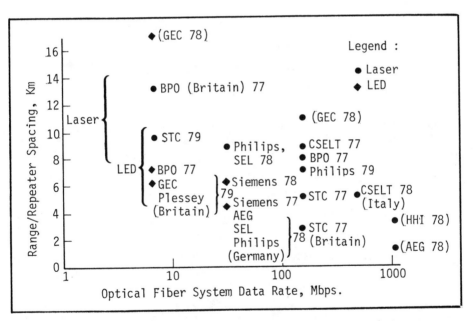

Figure 10.1 - First Generation European Optical Fiber Systems from 1977 to 1979. (Source: J. Midwinter, IEEE Spectrum, March 1980).

10.4 CURRENT AND PROJECTED COSTS

From various conferences, it has become evident that industry is ahead of predictions in installing optical fibers for communications. This is an encouraging sign, since greater production will bring the cost down more rapidly, and lowered prices will stimulate a broadening of the market and an increase in the types of applications for optical fibers.

10.4.1 Value-in-Use

Regarding the cost of using fiber optic technology in a product, it is important to determine the value of its unique assets to the customer. This means that for a specific application of fiber optics, a dollar value must be placed on all parameters constituting the value-in-use (VIU), which is given by the following formula.[7]

$$VIU = L_f(\Delta V_o + \Delta F_o/m) - F_{If}/m - V_{If} + (L_f/L_c)(V_c + V_{Ic} + F_{Ic}/m) \qquad (10.1)$$

VIU = Fiber optic cable value-in-use $\left(\$/m\right)$.

L_f, L_c = Life expectancy of the fiber optic and conventional systems, respectively $\left(years\right)$.

V_c = Selling price of conventional cable $\left(\$/m\right)$.

ΔV_o = Difference between all the variable operating costs of the fiber optic and conventional systems ($/m/year).

ΔF_o = Difference between all the fixed operating costs of the fiber optic and conventional systems ($/year).

m = Data transmission distance $\left(m\right)$.

F_{If} = Fixed investment cost of a fiber optic system $\left(\$\right)$.

F_{Ic} = Fixed investment cost of a conventional system $\left(\$\right)$.

V_{If} = Variable investment cost of a fiber optic system $\left(\$/m\right)$.

V_{Ic} = Variable investment cost of a conventional system $\left(\$/m\right)$.

The above formula is a simple formula, but holds only for certain applications. More information about various adjustments that must be made for wider use of this formula is given in Reference 7.

10.4.2 Cost Trends

When compared with other transmission media, optical fiber transmission costs less for trunking around 1000 two-way voice-grade circuits, as shown in Fig. 10.2. Twisted-pair and coaxial cables are
still more economical at lower and higher data rates, respectively.[2]

Optical fiber prices should continue to fall rapidly. This is
shown in Fig. 10.3 which represents an estimate presented to the
Electronic Industries Association by GTE's E.B. Carne.[2,9]

Continued cost reduction will be essential. It must be noted,
however, that many of the results of cost reduction programs usually
are not reported except on price sheets. As a representative report
on cost reduction, Fig. 10.4 shows continued lower prices and improved
performance of optical fibers from 1975 to 1980, as reflected in
Corning's price lists.[8] At the high performance end, prices have
dropped by a factor of three, and by approximately a factor of ten in
the low performance end.

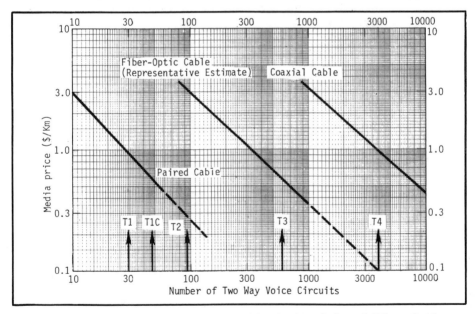

Figure 10.2 - Cost Comparisons for Paired, Coaxial and Fiber Optic
Cables (*After Reference 2*).

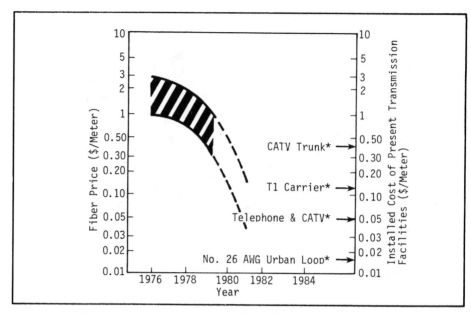

Figure 10.3 - Optical Fiber Prices Should Continue to Fall Rapidly. Estimates Presented to the Electronic Industries Association by GTE's E.B. Carne Show Fibers soon Becoming Competitive for T1 Carrier Systems (*After Reference 9*).

Although fiber optic prices continue to drop, operating systems may not achieve continuous price reductions. This is because cabling and installation, both constituting a large part of the total system cost, are traditional technologies that have already been extensively reduced in cost. The challenge for the optical fiber industry is to optimize the contribution of electronic and fiber costs to the whole system cost in order to benefit the entire telecommunications industry.

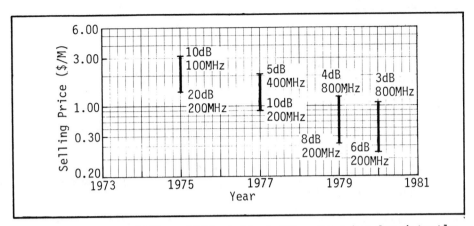

Figure 10.4 - From 1975 to 1980, Optical Fiber Cost has Consistently gone down while Performance had gone up, as Shown by Corning Glass Works Prices *(After Reference 8)*.

10.5 REFERENCES

1. Yancy, G.H.B.; *Fiber Optic Cables Increase Efficiency of Digital Transmission System;* Computer, pp. 59-64, November 1980.

2. Adlerstein, S.; *Message for the 80s: Technologies Mix to Bind Tele & Data Comm;* Electronic Design 1, pp. 86-89, January 4, 1980.

3. Fleming, J.W.; *High Rate Lightguide Fabrication Technique;* American Ceramic Society, 82nd Annual Meeting, Chicago, Il., April 27-30, 1980.

4. Williamson, J.; *Fiber Optics - The Light Fantastic;* Communications Engineering International, p. 33, August 1980.

5. Shimada, S; *Optical Systems: A Review - II. Japan: Unusual Applications;* IEEE Spectrum, pp. 74-76, October 1979.

6. Midwinter, J.E.; *Optical Fiber Systems: Low-Loss Links;* IEEE Spectrum, pp. 45-47, March 1980.

7. Uradnisheck, J.; *Estimating When Fiber Optics Will Offer Greater Value-in-Use;* Electronics, pp. 118-124, November 9, 1978.

8. Charlton, D. and Schultz, P.C.; *Progress in Optical Waveguide Processes, 1980;* Electro-Optical Systems Design, pp. 23-29, December 1980.

9. Carne, E. Bryan; *Telecommunications Trends: Fiber Optics;* Presented to the Electronic Industries Association, 1979.

CHAPTER 11

FUTURE TRENDS

Fiber optics will displace both twisted-wire-pair and coaxial cables in numerous existing applications and will make possible new applications requiring higher transmission quality. However, the prediction of the increase in optical fiber systems in the market place depends less on technical performance than on the cost reductions achievable while maintaining a high technical performance.

In this chapter, comments and forecasts are presented in four broad categories: technical parameters, applications, cost and market size.

11.1 INTRODUCTION

It is of interest to consider briefly what developments and system realizations may appear in the future, and which are already the subject of advanced research. Systems with optical signal processing can be regarded as *second generation*.

Future systems for home and business will integrate communications, control, surveillance and information retrieval. These systems will need the unique capabilities of optical cables to provide broadband, low attenuation and compatibility with a variety of signal formats and data encoding schemes. In fact, it is difficult to envision such systems without optical cables.

As far as cost-performance consideration is concerned, optical cables will displace mature copper conductor technology when one of two things happens: (1) optical cables do the same job or better at lower cost; (2) optical cables perform a task that conventional cables cannot perform well or at all.

Very little controversy now remains about the fact that fiber optic component production will grow rapidly to develop into a major worldwide market.

11.2 FIBER OPTICS IN FUTURE TELECOMMUNICATIONS

Based on present knowledge, the fiber optic communication systems of the future will consume very little power and will be of compact size. Still, the transmission of information will include data for television, telephones, and computers for more than 20 Km between repeaters and signal reconstruction systems at 1 Gbps or faster. These systems will most likely use emitters and detectors operating between 1.0 and 1.6 µm.

11.2.1 Technical Parameters and Cable Cost in the 80's

Table 11.1 presents some prognostic data on fiber optic technical parameters. As can be seen from this table, fibers with low loss and wide bandwidth will be available for operation over a spectral range wide enough to allow wavelength multiplexing.

Table 11.1 - Technical Parameters[1]

Fiber Loss	High quality, top of the line single-mode fibers will have a loss of 0.008 dB/Km. This primary loss reduction will have resulted from improvements in optical material purity. Significant improvement will also have been made possible by the advent of crystalline structures for fibers. Multimode fibers will have reached 0.02 dB/Km.
Splice Loss	Splice losses for field compatible applications will be in the 0.05-0.1 dB range.
Fiber Length	Routinely, unspliced lengths for communication fibers will be 5 Km. Lengths approaching 15 Km will be produced occasionally for special applications.
Fiber Bandwidth	Single-mode fiber systems will be operating on 2-Km links at 30 GHz with a 10 GHz information bandwidth. Multimode systems will have a 1 GHz bandwidth centered at 6 GHz and will routinely operate over 1 Km links.

One of the most difficult problems is to attain very high material purity during the manufacturing of fiber optics. Melting glass in space for later fiber drawing on earth may be one way to process ultrapure fiber materials. Such ultrapure materials are nearly

impossible to produce on earth because the glass often becomes con-
taminated during melting in a container. According to researchers of
Battelle's Columbus Laboratories, who are conducting ground-based experiments on ways to produce ultrapure glass preformed in space, the
contamination is caused by the container walls necessary for confinement during melting. However, this contamination could be alleviated in the microgravity of space, where melting can occur in container-
less environments.

Some cost indications that depend on cable features and type of
application are given in Table 11.2.

Optical fibers will be used for wide-band (1 GHz or greater),
two-way communication in nearly every household. Work centers with
wideband video and audio service will be located in or near homes to
reduce transportation costs. The electronic office will be connected
to the world via fibers. Optical fibers will interconnect the
cellular mobile radio telephone and will be used extensively in security systems, inventory control systems, and for communications in
transportation systems. RF and microwave communications will only be
used for mobile applications.

Table 11.2 - Cost Indications for the 80's[1]

0.1 dB/Km single mode fibers with rugged-ized protective jacket	Long haul buried cable	28¢ per meter in quantities of 500,000 meters (low-loss fiber portions will be 15% of the total)
Single unjacketed multi-mode fibers in the 10-20 dB/Km range		1¢ per meter in 500,000 meter quantities
Low loss fibers	Used to direct intensive laser energy at deu-terium pellets in fusion power plants.	A 1.5-meter long, 1.5-cm diameter core fiber will cost $12.5 K.

11.2.2 The Need for Higher Bandwidth

There is a very important technology impact that fiber optics
can make in communications and that is bandwidth. As the demand for
volumes of data continues, existing cable routes and ducts will be
stressed beyond capacity. The most cost-effective way of increasing
existing route data capacity is to substitute higher bandwidth

waveguides such as the optical fiber. However, it is foreseeable that even optical fibers carrying 45 million bits of data each second will soon be filled to capacity. Taking into consideration the newer high-bandwidth data terminals with video displays, extremely high bit rates between 300 million to 1 billion per second could be demanded.[2] With existing low-loss silica fibers, the 1 billion bit per second (1 Gbps) will be attained with faster light sources and detectors. The use of these components will be limited by the capability of data terminals to handle data at that high rate.

An even further look into the future might reveal the use of the integrated optical circuit, similar in philosophy to the electronic integrated circuit, with encoding, modulation, emission/detection components, and a fiber pigtail on a small square chip.

11.2.3 Digital Telephone Trunks

The future of fiber optic systems in high-density digital telephone trunks is quite bright. All new and replacement high-density digital transmission cable systems may eventually be implemented using fiber optic technology because of the greater capacity in less physical space at a potentially lower cost. Unlike copper cable, the price of fiber optic cable is declining. Similarly, as large scale integration technology matures, the cost of the multiplexing electronics required to take full advantage of fiber optic transmission will also decline.[3] Fiber optic technology is, in fact, a revolutionary advancement in data processing and communications electronics. Practical, economically viable systems exist today for many high-density digital trunking applications. In the future, this technology will not only meet presently conceived applications, but will undoubtedly stimulate the generation of additional applications and application concepts.

11.2.4 Local-Loop Networks

For fiber optic communications in local-loop networks, the future is also excellent. However, it will take until 1985 for an initial production system before costs are comparable to existing technologies. Still, a single local network designed to carry voice, video, and data should be realized in the next 15 years.[4]

In computer systems, fiber optics offers the greatest potential to replace the cumbersome parallel buses used in most of them.

11.2.5 Optical Communications vs. Microwave Transmission Systems

In many cases, optical communications versus microwave appears more complementary than competitive. In fact, the competition stands mainly with still existing techniques such as copper cables, and VHF-UHF radio.

Optics will be the background of future multi-service networks, which will need capacity transmission. Microwave transmission will predominate in rural areas, where low or medium capacity is needed and the cost of cable installation is high. In a second stage these equipments can be replaced by optical cable and re-used on other sites owing to their flexibility.

11.3 SOME SPECIAL APPLICATIONS AND TRENDS

As optical component performance continues to improve, new
equipment and systems will be generated and new applications will
evolve. Many major companies, not now actively involved in fiber
optics, will enter the field over this decade, often by acquisition
of, or joint ventures with, earlier participants.

11.3.1 Transmission of Infrared Laser Power [6,7,8]

Hughes Research Labs, Ca., has developed a new optical fiber
which can be used as a flexible waveguide which, unlike silica glass
fibers, is capable of transmitting several watts of infrared (IR)
laser power. Made from thallium bromo-iodide, this fiber optic system is designed to produce a 10-μm output of CO_2 laser. Also known
as KRS-5, the thallium bromo-iodide has the capability of transmitting infrared wavelengths to 25 μm.

The thallium bromo-iodide optical fibers contain a mosaic structure, in contrast to glass structures of conventional silica (SiO_2)
glass fibers. Even though silica fibers have been used traditionally
as waveguides for visible laser transmission, they do not transmit
in the longer wavelengths of the high-power CO_2 laser. Because the
new fiber system can transmit several watts of infrared laser power,
it is hoped that the use of laser surgery in dentistry, ophthalmology,
neurosurgery and urology will expand.

11.3.2 Automotive Applications

Because fiber optics provide immunity from EMI, with light weight
and elimination of ground loops, rapid growth is expected in automotive applications. Starting around 1983, the use of fiber optics for
the transmission of sensor and actuator data, as well as for dash-
board indicators, will become standard in the U.S. auto industry,
and will have spread to most car models in the 1985-1986 time frame.[9]
Since the distance which light must be transmitted in an auto is only
a few feet, the insertion loss through the connector is not a difficult requirement to meet. However, fiber optic connectors for automotive applications must be rugged, durable, and resistant to corrosion and vibration.

11.3.3 Undersea Optical Cable Systems[10,11]

Based on the belief that conventional coaxial technology has reached the end of the line, and on the expectation that within the next few years very long distance optical cable systems will be technologically and economically practicable, Bell is entering the applied research phase of an optical system that could greatly increase the capacity for trans-atlantic communications by the end of the decade. According to Dr. P.K. Punge of Bell Laboratories, undersea cables will continue to contribute significantly to meeting the evergrowing demand for international voice and data circuits by providing those circuits at less than the cost of satellite facilities. Using, for example, a cable with an outer diameter of 2.6 cm, containing up to 12 single-mode fibers of 10-μm core diameter, the system will require only about 200 repeaters to span the 6500-Km distance across the Atlantic Ocean. The expected circuit capacity would be 12,000 channels per pair of optical fibers. This corresponds to a digital transmission of 274 Mb/s and a bit rate reduction ratio of 3. Bell Laboratories looks toward the goal of an operational system that will provide digital transmission of voice, data and TV to Europe by 1990, with a future extension in a Y * configuration providing branches to, for example, Great Britain and the Continent.

11.3.4 All-Optical Telephones[12,13]

The feasibility of operating telephones solely by light transmitted over glass fibers has been demonstrated experimentally with a photodetector and transmitter developed by Bell Laboratories' researchers. In such an arrangement no self-contained battery, wall-outlet power, electricity or copper-wire connection from the telephone company office is required. Ringing is accomplished by photodetector conversion of laser light-pulses to electric power. When the phone is answered, electronic and optoelectronic circuits are connected, and the photodetector senses the pulse width modulated information in the laser signals. The resulting electric signals are converted to analog signals to drive the phone's receiver. The speaker's voice, converted to an amplitude modulated current pulse, causes the photodetector to function as an LED, and the diode transmits the pulse to the other end where it is converted back to an analog voice signal.

Work on the all-optical phone concept is still in the exploratory stages and considerable further work is required before such a phone can be considered for customer use.

---

* Y configuration = A star network of three branches.

11.3.5 Medical Applications

In the medical field, fibers will be used to carry signals from regulating devices to the nerve endings in the heart. This application will significantly reduce susceptibility of regulating devices to electromagnetic interference. Elastic fibers will play a key part in artificial nerves, and will also be used to carry data to and from prosthetic integrated circuits implanted in human bodies.

11.4 FUTURE FIBER OPTIC MARKET TRENDS

Based on various published market surveys, world-wide utilization of fiber optic systems will expand dramatically in the 1980's with the telecommunications industry being the largest consumer of fiber optics and related components.

11.4.1 Earlier Forecasts for U.S. Fiber Optics Growing Market

Forecasts prepared by various organizations, during the years 1976-1978 for the US Fiber Optic Component Market in 1985, are summarized in Fig. 11.1. These forecasts exclude systems value added and electronic hardware.[14]

The consensus of these forecasts is that economic justification for fiber optics is no longer in question for high capacity links, and in links where the particular technical attributes of fiber optics come into play.

Since fiber optics is a very dynamic field, such forecasts and predictions will be adjusted or revised depending on the course of fiber optic technology.

11.4.2 Fiber Optic Demand Flow

While fiber optic systems will be used in a significant number of different application areas, telecommunications applications will be the dominant consumer of this product. To arrive at this conclusion, the overall fiber optic demand flow has been modeled utilizing a combination of econometric and input/output matrix analysis techniques.[15] The fiber optic demand flow is shown in block diagram form in Fig. 11.2. In this demand flow, an optical fiber communication system has been defined to include the electro-optic transmitter, connectors, cable, repeaters, and electro-optic receiver. It

11.8

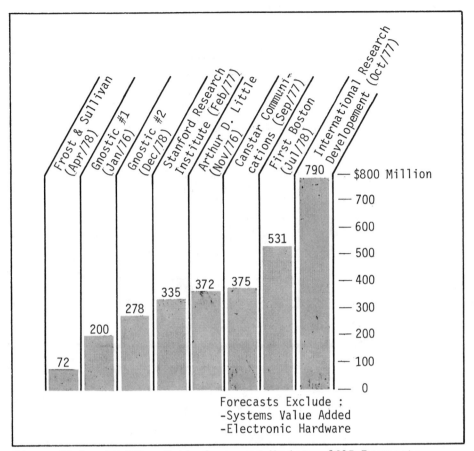

Figure 11.1 - US Fiber Optic Component Market - 1985 Forecasts.

specifically excludes any other electronics which precede or succeed
the electro-optic transmitter and receiver such as multiplexers or
demultiplexers. It does include the value of the electronics required to drive the transmitter and receiver, but excludes additional
circuitry, such as is required to provide standard alarm and protection functions.

11.4.3 Recent U.S. Fiber Optic Market Forecasts*

US production of communication fiber optic components in 1970,
exlucing R&D, was less than $1 million. This expanded to $4 million

* Most of the material of subsections 11.4.3 - 11.4.5 are excerpts
from Ref. 16 with permission of the authors.

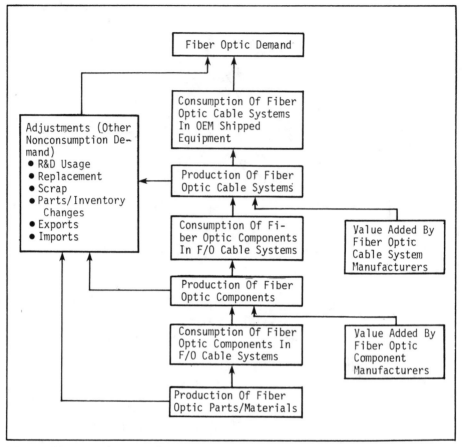

Figure 11.2 - Fiber Optic Demand Flow *(After Reference 15)*.

in 1975, and reached $50 million in 1980. The production of fiber
optic components will expand explosively over the next five years,
with an average growth rate of 49 percent per year, reaching $375
million* in 1985, in spite of a continuing rapid fall in unit prices.
Beyond 1985 through the end of this century, growth will continue im-
pressively upward, although at a gradually slowing rate. The US
market for fiber optic related programs will expand from $145 million
in 1980 to $3.48 billion in 1990. The largest share of this, 47 per-
cent in 1990, will be fiber optic cable systems, with the balance
consisting of interface electronics and planning/installation costs
as shown in Fig. 11.3.

---

* All values are in current dollars, and include a forecasted 8 percent average inflation rate 1980-1990.

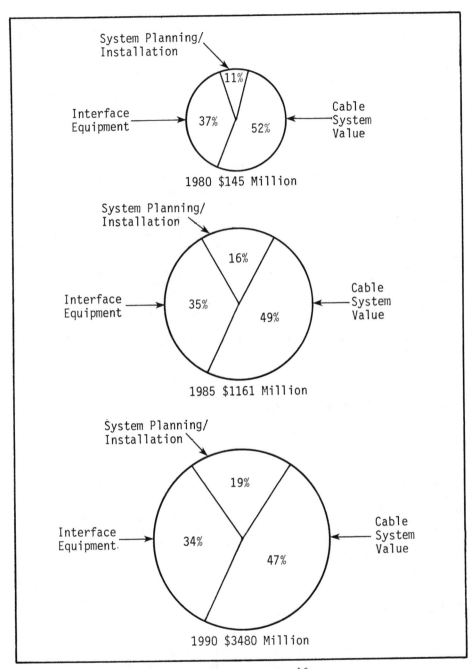

Figure 11.3 - US Fiber Optic Program Value[16]

US production of fiber optic components alone will climb from $50 million in 1980 to $1.06 billion in 1990. The largest portion of this market will be held by cable production, reaching a 57 percent share, $606 million, in 1990. Active circuit modules, including transmitters, receivers and repeaters, will hold the next largest market share. Connectors, couplers and miscellaneous items will gain a 12 percent share, reaching $127 million in 1990. It should be noted that this includes only cable termination connectors; there will be a comparable market for connectors used in fabricating active modules and other components.

Currently, military and other government applications represent the largest system end use, 37 percent of total consumption. Commercial telecommunications, however, will expand its fiber optic usage very rapidly. The Northeast corridor link, recently announced by AT&T, is only the first of many long-haul links to be installed by AT&T through the 80s. Underseas cable systems will become significant in the late 1980s. Individual telephone operating companies are moving aggressively into fiber optics use. Substantial installations already are planned for 1980 through 1983, and much larger programs are now in preliminary planning. By the mid-1980s, fiber optics will become a substantial factor in subscriber broadband network connection. Taking all these factors into account, the commercial telecommunication use of fiber optic components will expand to $458 million or 43 percent of the total, in 1990. Commercial computer systems, industrial, automotive and instrumentation applications, however, also will become worthwhile markets.

11.4.4 World Market Potential

Fiber optic component development and application has advanced in Japan and in numerous European countries, essentially in parallel with US efforts. The total free world production of fiber optic components in signal transmission applications will expand dramatically from $85 million in 1980 to $1.86 billion in 1990. North American production will hold the largest share, as shown in Fig. 11.4, climbing to $1.16 billion or a 62 percent share by 1990.[16] A substantial share of component production, especially in Japan and Europe, will be consumed by domestically produced communication systems which will be exported to third world countries.

11.4.5 Long-Range Potential

Beyond 1990, nearly all new signal transmission lines more than a few meters in length (and many that are shorter) will be fiber optics. The growth rate of the fiber optic-related program production

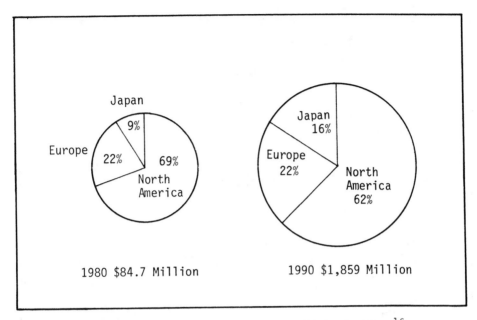

Figure 11.4 - Fiber Optic Component Free World Production.[16]

value will continue to slow over the 1990-2000 decade, but will still continue at a stronger pace than the overall world electronic industry. The free world production cost of communication fiber optic programs will climb from $275 million in 1980 to $6.5 billion in 1990 and to $40 billion in 2000,[16] as noted in Fig. 11.5. Cable systems will continue as the dominant program cost element.

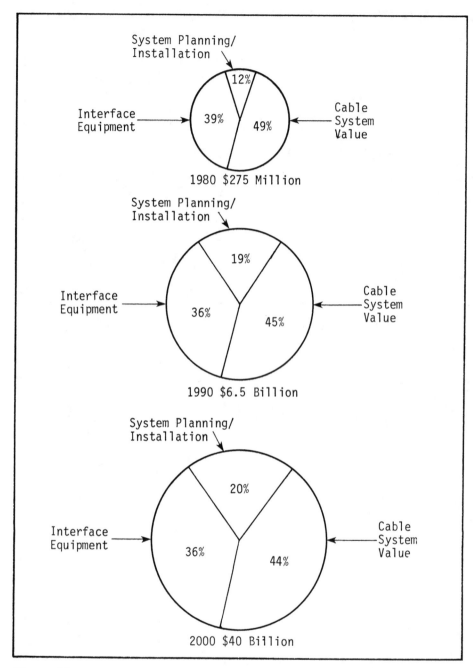

Figure 11.5 - Fiber Optic Program Worldwide Demand.[16]

11.5 REFERENCES

1. Beiser, R.L.; *Fiber Optics in 1989;* Electro-Optical Systems Design, pp. 61-62, July 1979.

2. Klein, J.R.; *Fiber Optic Light Source;* Telecommunications, pp. 45-46, September 1979.

3. Yancy, G.H.B.; *Fiber Optic Cables Increase Efficiency of Digital Transmission Systems;* Computer Design, pp. 59-64, November 1980.

4. Heftman, G.; *Fiber Optics Debuts in Telecomm, but Lags in Datacomm;* Electronic Design, p.40, September 1, 1980.

5. Dupuis, Ph. and Threheux, M.; *Optical Communications Versus Microwave Transmission Systems;* 10th European Microwave Conference, Warsaw, Poland, 8-12 September 1980.

6. Late News; *Fibers Handle CO_2-Laser Wavelengths;* Electronic Design, p. 23, July 19, 1980.

7. U.S. News; *New Hughes Fiber Optics System Transmits Several Watts of Infrared Laser Power;* Fiber Optics and Communications, Vol. 3, No. 9, pp. 4-5, September 1980.

8. Robertson, M.M.; *Power Transfer via Fiber Optics;* 30th Electronic Components Conference, San Francisco, Ca., April 28-30, 1980.

9. Technology Trends; *Automotive Markets Take Off;* Digital Design, p. 16, April 1980.

10. News; *Bell Proposes Undersea Optical Cable System;* Electro-Optical Systems Design, p. 12, May 1980.

11. Runge, P.K.; *High-Capacity, Optical-Fiber Undersea Cable System;* CLEOS, San Diego, Ca., Feb. 26-28, 1980.

12. Technology News; *Experimental Fiber-Optic Telephone Requires no External Power;* EDN, p. 44, January 5, 1979.

13. Miller, R.C. and Lawry, R.B.; *Optically Powered Speech Communication Over a Fiber Lightguide;* The Bell System Journal, Vol. 58, No. 7, pp. 1735-1741, September 1979.

14.   Canstar; *Canstar Forecasts Strong, Growing Fiber Optics
       Market;* The Canstar Newsletter of Fiber Optics Technology,
       Vol. 2, No. 2, p. 3, Summer 1979.

15.   Strachman, H.L.; *Worldwide Telecommunications Utilization
       of Fiber Optics;* 3rd World Telecommunication Forum, TELECOMP
       79, Geneva, Switzerland, September 24-26, 1979.

16.   Montgomery, J.D. and Dixon, F.W.; *Fiber Optic Market Trends;*
       12th Annual Electro-Optics/Laser 80, Boston, Ma., November
       19-21, 1980.

INDEX

A

B

C

Index

Index

Index

Index

M

N

O

Index

Index